AI-Enhanced Operation and Management of Renewable Energy-Integrated Power Systems

AI-Enhanced Operation and Management of Renewable Energy-Integrated Power Systems

Guest Editors

Jiaqi Ruan
Yujia Huang
Chao Yang

MDPI

Basel • Beijing • Wuhan • Barcelona • Belgrade • Novi Sad • Cluj • Manchester

Guest Editors
Jiaqi Ruan
College of Electrical Engineering
Sichuan University
Chengdu
China

Yujia Huang
College of Electrical Engineering
Shenyang University of Technology
Shenyang
China

Chao Yang
School of Electrical and Electronic Engineering
North China Electric Power University
Baoding
China

Editorial Office
MDPI AG
Grosspeteranlage 5
4052 Basel, Switzerland

This is a reprint of the Special Issue, published open access by the journal *Energies* (ISSN 1996-1073), freely accessible at: https://www.mdpi.com/journal/energies/special_issues/914V3157FG.

For citation purposes, cite each article independently as indicated on the article page online and as indicated below:

Lastname, A.A.; Lastname, B.B. Article Title. *Journal Name* **Year**, *Volume Number*, Page Range.

ISBN 978-3-7258-7090-5 (Hbk)
ISBN 978-3-7258-7091-2 (PDF)
https://doi.org/10.3390/books978-3-7258-7091-2

Contents

About the Editors

Jiaqi Ruan

Jiaqi Ruan is an associate professor in the College of Electrical Engineering at Sichuan University, Chengdu, China. His research focuses on smart grids and renewable-rich power systems, with particular interests in AI-enabled operation and management, cyber–physical security, climate-related risk, and vulnerability analysis. His work addresses the interaction between advanced data-driven methods and the physical and operational constraints of modern power systems. He has published over 50 SCI- and EI-indexed journal papers, including articles in *Nature Reviews Electrical Engineering, The Innovation,* and *Nexus,* and one of his papers was recognized as a Popular Paper in *IEEE Transactions on Smart Grid.* He has led and participated in multiple competitive research projects, including grants supported by the National Natural Science Foundation of China. In addition to his research activities, he has served as a session chair and invited speaker at several international conferences and is a member of the Youth Editorial Board of *Protection and Control of Modern Power Systems and Energy Conversion and Economics.*

Yujia Huang

Yujia Huang received bachelor's, master's and Ph.D. degrees from the Northeastern University, China, in 2018, 2020, and 2024, respectively. She is currently a Lecturer with the Department of Electrical Engineering, Shenyang University of Technology, China. Her main research interests include modeling and analysis of distributed resources and integrated energy systems. She has published over 20 papers in journals such as *IEEE Transactions.* Two papers published by the first author have been recognized as ESI Highly Cited Papers (Top 1%). She received the Outstanding Thesis Award from *IEEE Smart Grid Comm* in 2025, Best Paper Awards form *ICEI* in 2022, and *IEEE EI2* in 2025. She was selected for the First Liaoning Province Young Science and Technology Talents Support Project.

Chao Yang

Chao Yang is a lecturer at the School of Electrical and Electronic Engineering, North China Electric Power University, China. He earned his Ph.D. degree in Engineering from the North China Electric Power University and subsequently joined the Chinese University of Hong Kong (Shenzhen) as a postdoctoral researcher. His research interests focus on the electricity market, low-carbon electricity and AI/LLM for power systems.

Preface

The rapid growth of renewable energy penetration is fundamentally reshaping the operation and management of modern power systems. Increasing variability, uncertainty, and system coupling challenge traditional operation paradigms that rely heavily on deterministic models and centralized decision-making. At the same time, recent advances in AI have created new opportunities to enhance situational awareness, forecasting accuracy, and operational adaptability across multiple temporal and spatial scales.

This Reprint was motivated by the growing recognition that AI is no longer a peripheral tool, but an increasingly integral component of renewable-rich power system operation and management. Rather than focusing on isolated algorithms, the contributions collected here emphasize how data-driven and learning-based methods can be embedded into practical operational workflows, including forecasting, state estimation, scheduling, optimization, and coordinated control. The included articles span a range of system contexts, from active distribution networks and integrated energy systems to coupled infrastructures, reflecting the expanding scope of power system operation under high renewable integration.

The aim of this Reprint is to provide a coherent overview of recent research efforts that bridge AI techniques with the physical, operational, and market constraints of real-world power systems. By highlighting both methodological advances and application-oriented studies, this Reprint seeks to support researchers, engineers, and decision-makers who are interested in developing reliable, efficient, and low-carbon operational strategies for power systems undergoing rapid transformation. It is intended as a reference for those exploring how AI-enhanced approaches can contribute to the next generation of renewable-rich power systems

Jiaqi Ruan, Yujia Huang, and Chao Yang
Guest Editors

energies

MDPI

Article

Dynamic Reconfiguration Method of Active Distribution Networks Based on Graph Attention Network Reinforcement Learning

Chen Guo, Changxu Jiang * and Chenxi Liu

College of Electrical Engineering and Automation, Fuzhou University, Fuzhou 350108, China; 230127034@fzu.edu.cn (C.G.); 220120055@fzu.edu.cn (C.L.)
* Correspondence: cxjiang@fzu.edu.cn or jiang316300541@126.com

Abstract: The quantity of wind and photovoltaic power-based distributed generators (DGs) is continually rising within the distribution network, presenting obstacles to its safe, steady, and cost-effective functioning. Active distribution network dynamic reconfiguration (ADNDR) improves the consumption rate of renewable energy, reduces line losses, and optimizes voltage quality by optimizing the distribution network structure. Despite being formulated as a highly dimensional and combinatorial nonconvex stochastic programming task, conventional model-based solvers often suffer from computational inefficiency and approximation errors, whereas population-based search methods frequently exhibit premature convergence to suboptimal solutions. Moreover, when dealing with high-dimensional ADNDR problems, these algorithms often face modeling difficulties due to their large scale. Deep reinforcement learning algorithms can effectively solve the problems above. Therefore, by combining the graph attention network (GAT) with the deep deterministic policy gradient (DDPG) algorithm, a method based on the graph attention network deep deterministic policy gradient (GATDDPG) algorithm is proposed to online solve the ADNDR problem with the uncertain outputs of DGs and loads. Firstly, considering the uncertainty in distributed power generation outputs and loads, a nonlinear stochastic optimization mathematical model for ADNDR is constructed. Secondly, to mitigate the dimensionality of the decision space in ADNDR, a cyclic topology encoding mechanism is implemented, which leverages graph-theoretic principles to reformulate the grid infrastructure as an adaptive structural mapping characterized by time-varying node–edge interactions Furthermore, the GATDDPG method proposed in this paper is used to solve the ADNDR problem. The GAT is employed to extract characteristics pertaining to the distribution network state, while the DDPG serves the purpose of enhancing the process of reconfiguration decision-making. This collaboration aims to ensure the safe, stable, and cost-effective operation of the distribution network. Finally, we verified the effectiveness of our method using an enhanced IEEE 33-bus power system model. The outcomes of the simulations demonstrate its capacity to significantly enhance the economic performance and stability of the distribution network, thereby affirming the proposed method's effectiveness in this study.

Keywords: active distribution network; dynamic reconfiguration; graph attention network; deep deterministic policy gradient; deep reinforcement learning

https://doi.org/10.3390/en18082080

1. Introduction

In September 2020, China outlined its aspirations for carbon peaking and achieving carbon neutrality. DGs typically use renewable energy as a source of energy, which can effectively reduce carbon emissions and contribute to implementing the "double carbon policy" goals. Hence, numerous distributed generations, such as wind turbines and solar photovoltaic systems, have been incorporated into the electricity grid. Consequently, there is a growing transition away from traditional distribution systems towards active distribution networks. Nonetheless, the variability and unpredictability in DG production can result in heightened power wastage and voltage fluctuations, ultimately impacting the reliability of the electricity supply. This creates difficulties for the economic efficiency, dependability, and operational steadiness of the electrical network [1]. ADNDR enhances the efficiency of the distribution network's structure and function through the manipulation of contact switches, either by opening or closing them. This approach significantly boosts the share of renewable energy utilized and accelerates the transformation and upgrading of the energy system [2]. It is a very effective technology for distribution network management and operation. It alters the operational status of switches to reconfigure the network layout during standard operation of the electricity distribution system, fulfilling the objectives of minimizing network losses and enhancing voltage regulation. At present, some reconfiguration strategies involve one-time static reconfiguration of the distribution network, but static reconfiguration cannot respond to load changes promptly. Dynamic reconfiguration can dynamically manage and optimize the power grid to improve its flexibility and response speed [3]. ADNDR represents a sophisticated, multi-dimensional, mixed-integer, nonlinear optimization challenge characterized by stochasticity. The scale of this problem increases sharply with the number of distribution network branches. Hence, there is an immediate necessity to explore a highly effective model and technique for swiftly and precisely computing the ADNDR approach [4].

In Reference [5], the mathematical framework of distribution network reconfiguration is addressed through the application of second-order cone approximation, which is then reformulated as a mixed-integer second-order cone programming (MISOCP) challenge and resolved. Although this method can accurately find the optimal reconfiguration strategy for small-scale distribution networks in a short period, the MISOCP method is difficult to model, computationally intensive, and requires precise distribution network parameters for model construction. Therefore, using this method to solve ADNDR problems in large-scale distribution networks requires a long time. Reference [6] proposes an ADNDR strategy based on the retained branch exchange (RBE) algorithm to reduce network losses in distribution networks. Although the RBE algorithm outperforms mathematical optimization algorithms in computational efficiency, it has a high dependence on the initial topology of the distribution network, which makes it prone to falling into local optima. In Reference [7], a solution to the problem is proposed through an ADNDR strategy utilizing a combined particle swarm optimization (PSO) approach. This solution combines the advantages of the binary PSO algorithm and the traditional PSO algorithm and has better optimization performance compared to other algorithms. Nevertheless, the PSO algorithm primarily depends on localized data and adjacent searches, making it susceptible to becoming trapped in suboptimal solutions.

Despite yielding promising outcomes, the aforementioned techniques all exhibit drawbacks, including lengthy computation periods, the necessity for comprehensive distribution network parameters, and a tendency towards converging to local optima. Reference [8] introduces a distribution network reconfiguration approach grounded in deep reinforcement learning (DRL). This strategy does not necessitate comprehensive distribution network parameters and is capable of significantly minimizing network losses and voltage deviations. Nevertheless, the methodology employs Convolutional Neural Networks (CNNs) to estimate the action value function, neglecting the graph structure information of the distribution network. Consequently, there is room for enhancement in its optimization outcomes. Reference [9] proposes a distribution network reconfiguration strategy based on graph reinforcement learning. This strategy uses graph convolutional networks (GCNs) to fit the action value function and learns complex connection patterns between nodes through capsule structures and graph neural networks to obtain better network reconfiguration solutions.

The distribution network fundamentally possesses a graphical model structure. Currently, a significant portion of the literature on ADNDR strategy research lacks consideration for the power grid's topology. Unlike traditional data structures, graph models frequently utilize irregular, non-Euclidean structured data for representation. Furthermore, graph neural networks, leveraging their strengths in handling graph-based information, offer significant support in extracting features from power grid structures. But in the graph convolutional network (GCN), the weights between nodes are fixed and cannot adaptively learn the importance between different nodes. This will result in performance limitations for the GCN when dealing with complex graph data in distribution network structures. In addition, the node connections in the distribution network are relatively sparse, so the distribution network structure is usually a sparse graph, and the GCN needs to perform a fully connected operation on the entire graph, which will result in significant computational and storage costs when processing sparse graphs. The GAT incorporates an attention framework, enabling the model to discern the importance of connections among various nodes. Consequently, it can capture intricate relationships between nodes with greater adaptability. Moreover, because only the relationship between adjacent nodes needs to be considered when calculating attention weights between nodes, and because there is no need to perform fully connected operations on all nodes, the GAT is more efficient in processing sparse graph data similar to distribution network structures. In summary, compared to the GCN, using the GAT as a function-fitting tool for distribution network reconfiguration has better performance [10].

This paper introduces an innovative method, termed the ADNDR method, which builds upon the foundation of the GATDDPG algorithm. The newly introduced method leverages the strengths of the GAT alongside deep deterministic policy gradients, addressing the challenge of distribution network reconfiguration. In this way, it integrates their benefits effectively. By incorporating the GAT, the algorithm is able to precisely capture the topological and electrical properties of the distribution network. Consequently, it gains a deeper understanding of the system's operational status and identifies potential risks more effectively. Meanwhile, utilizing DDPG to optimize the reconfiguration decision process ensures that the best reconfiguration solution is found in the action space [11]. Consequently, the introduced dynamic reconfiguration approach for active distribution networks, leveraging GATDDPG, enhances the efficiency and precision of algorithms in managing intricate distribution systems. Additionally, it offers a novel framework to address distribution network reconfiguration challenges. The findings reveal that the method substantially enhances the reliability of the power system by smartly altering the distribution network's topology. Furthermore, it decreases operating costs and fosters the incorporation and utilization of renewable energy sources.

2. A Mathematical Model for ADNDR

2.1. Objective Function of ADNDR

To improve the voltage and network loss of the distribution network, the sum of the changes in network loss and voltage offset before and after reconfiguration is utilized to serve as the objective function for ADNDR.

$$f_1 = \sum_{t=1}^{T} P_{\text{loss},t} = \sum_{t=1}^{T}\sum_{l=1}^{N_l} x_{l,t} R_l I_{l,t}^2 = \sum_{t=1}^{T}\sum_{l=1}^{N_l} x_{l,t} R_l \frac{P_{l,t}^2 + Q_{l,t}^2}{U_{l,t}^2} \tag{1}$$

$$f_2 = \sum_{t=1}^{T} \Delta U_t = \sum_{t=1}^{T}\sum_{i=1}^{n} \left| \frac{U'_{i,t} - U_{\text{N}}}{U_{\text{N}}} \right| \tag{2}$$

where T is the simulation step size. In this paper, T is set to 24 h, meaning that each calculation step corresponds to a duration of 1 h. Therefore, the simulation progresses in hourly increments, allowing for detailed analysis of the network state at each time step. $P_{\text{loss},t}$ indicates the network loss incurred by the distribution network at time t, prior to any reconfiguration; ΔU_t represents the voltage offset of node i at time t before reconfiguration; N_l is the number of branches in the distribution network; and $x_{l,t}$ represents the state branch l at time t. When $x_{l,t}$ = 0, it indicates that the branch is open, and when $x_{l,t}$ = 1, it indicates that the branch is closed. R_l represents the resistance of branch l; $I_{l,t}$ represents the current of branch l at time t; $P_{l,t}$ and $Q_{l,t}$, respectively, represent the active and reactive power of branch l at time t; $U_{l,t}$ represents the terminal node voltage of branch l at time t; $U_{i,t}$ is the voltage of node i at time t; and U_{N} is the rated voltage.

Due to the different orders and dimensions of network loss and voltage offset, it is necessary to standardize them using max–min:

$$f_i' = \frac{f_i - f_{i,\min}}{f_{i,\max} - f_{i,\min}} (i = 1, 2) \tag{3}$$

where f_i' is the standardized objective function of f_i and $f_{i,\min}$ and $f_{i,\max}$ are the minimum and maximum values of f_i, respectively.

Therefore, the objective function of ADNDR is as follows:

$$\min f_{\text{ADNDR}} = \min \sum_{i=1}^{2} \lambda_1 f_i' \tag{4}$$

where λ_i is the weight coefficient of the i-th optimization objective f_i, and its specific value is determined through the Analytic Hierarchy Process [12].

2.2. Source Load Uncertainty Model for ADNDR

Multi-scenario technology is a method for describing uncertainty, and this section constructs an uncertainty output model for distribution network load demand and distributed renewable energy sources based on multi-scenario technology. Through multi-scenario technology, the model of the distribution network with source load uncertainty can be transformed into a deterministic scenario, simplifying the solution of the model.

Modeling the uncertain output of loads and renewable distributed power sources is performed as shown in Equation (5):

$$P_{\text{rand},i,t} = \hat{P}_{\text{rand},i,t} + \varphi_{\text{rand},i,t} \tag{5}$$

where $P_{\text{rand},i,t}$, $\hat{P}_{\text{rand},i,t}$, and $\varphi_{\text{rand},i,t}$ are the actual values, predicted values, and prediction errors of loads or renewable distributed power sources, respectively.

Assuming that the above errors follow a normal distribution with a mean of 0, the source load uncertainty output model can be generated using oversampling. In this paper, Latin Hypercube Sampling (LHS) is used to generate the load and renewable distributed power output scenarios. Compared with Monte Carlo simulation, LHS ensures that all sampling areas are covered by sampling points through hierarchical sampling. The sampling values of the LHS sampling random variables are:

$$P_n = F^{-1}(U_n) = F^{-1}[(U + n - 1)/N] \tag{6}$$

where P_n is the nth sampled value of variable P and F is the probability distribution function. The random number $U_n = (U + n - 1)/N$ represents the nth subinterval obtained by dividing the interval N of $[0, 1]$ into equal parts, where $U \in [0, 1]$.

By generating scenarios, the source load uncertainty scenarios of active distribution networks can be transformed into deterministic scenarios.

2.3. Constraints for ADNDR

The power flow Equations (7) and (8) constitute the fundamental equality constraints in distribution network analysis, representing the node power balance conditions that must be strictly satisfied during network state computation. Specifically, Equation (7) defines the active power balance constraint, while Equation (8) characterizes the reactive power balance constraint. Their mathematical expressions are formulated as:

$$P_{i,t} = U_{i,t}\sum_{j=1}^{n} U_{j,t}(G_{ij}\cos\theta_{ij,t} + B_{ij}\sin\theta_{ij,t}) = P_{\mathrm{DG},i,t} - P_{\mathrm{load},i,t} \tag{7}$$

$$Q_{i,t} = U_{i,t}\sum_{j=1}^{n} U_{j,t}(G_{ij}\sin\theta_{ij,t} - B_{ij}\cos\theta_{ij,t}) = Q_{\mathrm{DG},i,t} - Q_{\mathrm{load},i,t} \tag{8}$$

where $P_{i,t}$, represents the real power injected at node i at time t, while $Q_{i,t}$ signifies the reactive power injected at the same node and time. $P_{\mathrm{DG},i,t}$ and $Q_{\mathrm{DG},i,t}$ are the active and reactive outputs of the DG at node i at time t, respectively; $P_{\mathrm{load},i,t}$ and $Q_{\mathrm{load},i,t}$ are the active and reactive load power at node i at time t, respectively; $U_{i,t}$, $U_{j,t}$ are the node voltages of node i and node j at time t, respectively; G_{ij} and B_{ij} represent the conductivity and admittance between node i and node j, respectively; and $\theta_{ij,t}$ represents the phase angle difference in voltage between nodes i and j at a given instant t.

The distribution network's radial limitations restrict its maximum load-bearing capacity, and they also impact the level of node connectivity within the network. Therefore, radial constraints need to be considered when reconfiguring the distribution network. The radial constraint methods in distribution networks can be divided into five categories: (1) the power flow-based constraint method, (2) the virtual power flow-based constraint method, (3) the graph-based spanning tree constraint method, (4) the power supply path-based constraint method, and (5) the power supply loop-based constraint method. Among them, the most commonly used method is the power supply loop-based constraint method, which needs to satisfy two conditions. The one is that the network contains $N - N_s$ closed branches, and their mathematical representation is presented in Equation (9). The other is the absence of a connected power supply loop within the distribution network, as depicted in Equation (10).

$$\sum_{b=1}^{N_l} x_b = N - N_s \tag{9}$$

$$\sum_{m=1}^{M_l} x_{lm} \le M_l - 1 \tag{10}$$

where N_l signifies the quantity of branches present within the distribution network; x_b denotes the condition or status of the b-th branch, which is 1 when the branch is closed and 0 when it is open; N is the number of nodes in the network; N_s is the number of substations in the network; M_l is the number of branches in power supply loop l, $l = 1, 2, \cdots, L$, where L is the total number of power supply loops in the network; and x_{lm} is the state of the m-th branch m in the l-th power supply loop, with a value of 1 when the branch is closed and 0 when the branch is disconnected [13].

The stability of voltage amplitude is one of the important conditions to ensure the safety and normal operation of the power grid. If the voltage amplitude in the power grid lacks sufficient stability, it may lead to power equipment failure. Therefore, when reconfiguring the distribution network, node voltage constraints need to be considered, which are expressed as follows:

$$U_{\min} \le U_i \le U_{\max} \tag{11}$$

where the lower voltage threshold of the node is denoted as $U_{\min}$, while U_i represents the voltage magnitude at the i-th node. Additionally, the upper voltage threshold of the node is referred to as $U_{\max}$.

In the dynamic reconfiguration process of distribution networks, excessive switching operations not only accelerate performance degradation and shorten the service life of circuit breakers, but they also jeopardize system stability. Therefore, switching operation counts must be constrained as critical optimization objectives:

$$\begin{cases} \sum\limits_{t=1}^{T} \sum\limits_{b=1}^{N_{\text{SWI}}} \left| X_{b,t} - X_{b,t-1} \right| \le H_{\text{SWI,max}} \\ \sum\limits_{t=1}^{T} \left| X_{b,t} - X_{b,t-1} \right| \le H_{\text{SWI,b,max}} \end{cases} \tag{12}$$

where $H_{\text{SWI,max}}$ denotes the maximum allowable total switching operations for all switches involved in distribution network reconfiguration, N_{SWI} represents the number of operable switches in the distribution network, and $H_{\text{SWI,b,max}}$ specifies the individual switching operation limit per circuit breaker to prevent frequent switching of any single breaker, even when the total number of operations meets the overall requirement.

In the process of distribution network reconfiguration, it is essential to consider branch power constraints due to potential safety risks caused by uneven load distribution or line parameter discrepancies, which may lead to branch power exceeding limits, resulting in equipment overload and voltage instability [14]. The constraint is formulated as:

$$S_{t,l} \le S_{l,\max} \tag{13}$$

where $S_{t,l}$ is the apparent power of branch l at time t and $S_{l,\max}$ is the upper limit of the apparent power of branch l.

The mathematical framework presented in this paper for ADNDR encompasses two variables subject to uncertainty: distributed generation outputs and power loads. Therefore, the essence of the proposed reconfiguration mathematical model is a complex, nonlinear, stochastic optimization mathematical model.

3. Radial Topology Analysis of a Distribution Network

To efficiently decrease the quantity of viable solutions during the dynamic reconfiguration of active distribution networks and expedite the algorithm's execution, this paper employs a loop-centric encoding technique. This approach minimizes the action space within the network reconfiguration process.

The fundamental loop (FL) is characterized as the smallest circuit that excludes any other circuits within it. The IEEE 33-bus power system takes this as an example. There are five FLs, as shown in Figure 1. The branches within each FL are listed in Table 1.

Additionally, Table 2 presents the common branches found in the distribution network. Furthermore, Table 3 displays the remote control switches (RCSs) associated with each FL.

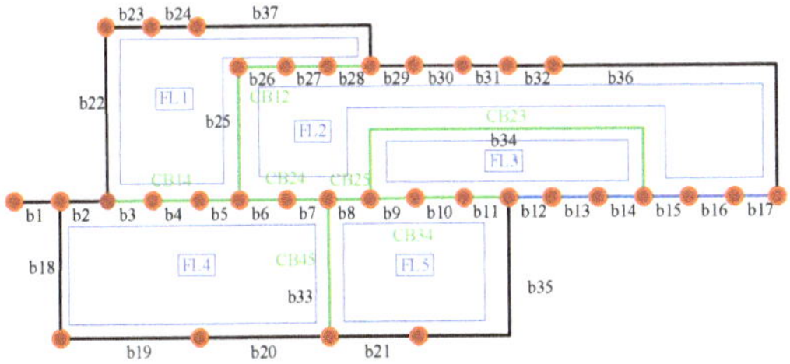

Figure 1. IEEE 33-bus power system branch encoding diagram.

Table 1. The fundamental loop branch set of the IEEE 33-bus power system.

Fundamental Loop	Branch Number
FL1	b28, b27, b26, b25, b5, b4, b3, b22, b23, b24, b37
FL2	b17, b16, b15, b34, b8, b7, b6, b25, b26, b27, b28, b29, b30, b31, b32, b36
FL3	b14, b13, b12, b11, b10, b9, b34
FL4	b20, b19, b18, b2, b3, b4, b5, b6, b7, b33
FL5	b21, b33, b8, b9, b10, b11, b35

Table 2. The common branches of the IEEE 33-bus power system.

Common Branch (CB)	Associated Branch	Branch Number
CB12	FL1 ∩ FL2	b25, b26, b27, b28
CB14	FL1 ∩ FL4	b3, b4, b5
CB23	FL2 ∩ FL3	b34
CB24	FL2 ∩ FL4	b6, b7
CB25	FL2 ∩ FL5	b8
CB35	FL3 ∩ FL5	b9, b10, b11
CB45	FL4 ∩ FL5	b33

Table 3. The remote control switches in each fundamental loop.

Fundamental Loop	RCSs
FL1	b3, b23, b27, b37
FL2	b7, b8, b27, b31, b34, b36
FL3	b9, b13, b34
FL4	b3, b7, b18, b33
FL5	b8, b9, b33, b35

When the CB of the two FLs is interrupted by more than one branch, an island will appear, causing the distribution network to malfunction. To avoid this situation, the following steps are proposed in this paper to obtain a feasible solution for the action space of branch switches in distribution network reconfiguration.

(1) List all CBs and RCSs.

(2) Generate an n-dimensional vector $Y = \{y_1, y_2, \ldots, y_n\}$ to store the final action space for distribution network reconfiguration, where n is the number of feasible solutions, y is a k-dimensional vector, and k is the number of FLs in the system. It means that each FL needs to be disconnected from a switch.

(3) Select a switch b_i in each FL to generate a k-dimensional vector x.

(4) Perform a feasibility check on each k-dimensional vector x, that is, check each combination of b_i and b_j ($i \leq n$, $j \leq n$, and $i \neq j$). If there is $b_i \in \mathrm{CB}_{ij}$ and $b_j \in \mathrm{CB}_{ij}$, it is removed from the final action space. If after traversing all b_i (i = 1, 2, . . ., n), if there is still no $b_i \in \mathrm{CB}_{ij}$ and $b_j \in \mathrm{CB}_{ij}$, then the feasible solution is retained.

It is worth noting that through step 3, there are no loops generated during the reconfiguration process of the distribution network. Furthermore, through step 4, it can be ensured that no two switches are in the same common branch in an action space, avoiding the occurrence of "islands" during ADNDR. Using the IEEE 33-bus power system as a case study, our encoding technique successfully decreases the dimensionality of the solution space by a remarkable 95.53%, from 2^{14} = 16,384 to 732 dimensions. In addition, the encoding method based on FLs can also ensure that all candidate solutions in the solution space meet the radial constraints of the distribution network. Therefore, this encoding method, which is based on loops, proves to be both effective and practical.

4. Graph Attention Network and Deep Deterministic Policy Gradient Algorithm

4.1. Graph Attention Network (GAT)

In this paper, the topology of the distribution network is described as an undirected network graph $\boldsymbol{G} = (\boldsymbol{V}, \boldsymbol{E})$, where $\boldsymbol{V}$ contains all N nodes in the distribution network, $v_i \in \boldsymbol{V}$, and $\boldsymbol{E}$ represents the edge between nodes, which is the line of the distribution grid, $(v_i, v_j) \in \boldsymbol{E}$. In addition, the topology structure of the distribution network is analyzed by using the undirected network diagram to represent the structural information. Based on this, to process the irregular non-Euclidean structured data, the GAT is adopted to capture complex relationships between nodes in the distribution network through graph methods, such as the connection methods between nodes, and learn different importance weights between nodes through attention mechanisms. The GAT algorithm excels at preserving the integrity of original data, minimizing information loss and distortion. Additionally, it effectively captures the intricate topology and electrical attributes within the distribution network. Consequently, it provides a deeper insight into the system's operational status and potential hazards. During the ADNDR process, the layout of the distribution network undergoes constant alterations. GATs possess a specific level of flexibility, enabling them to address these changes in the network's layout to a certain degree. This characteristic facilitates real-time surveillance of the distribution network and aids in devising reconfiguration strategies.

The input of the GAT is $\boldsymbol{I} = (\boldsymbol{G}, \boldsymbol{X})$, where $\mathbf{X} = (x_1, x_2, \ldots x_N) \in \mathbf{R}^{N \times C}$ is the eigenvector matrix of each node and $\boldsymbol{G} = (\boldsymbol{V}, \boldsymbol{E})$ is the undirected graph corresponding to the distribution network, where N is the number of nodes and C is the eigenvector dimension of each node. To improve the expression ability of each node's features, the GAT uses a self-attention mechanism for each node, with a self-attention coefficient of:

$$e_{ij} = a(\boldsymbol{W}x_i, \boldsymbol{W}x_j) \tag{14}$$

where a is a single-layer feedforward neural network; x_i and x_j are the feature vectors of node i and node j, respectively; and $\boldsymbol{W}$ is the weight matrix. Unlike the global attention mechanism, which considers all positional information of feature vectors, the GAT employs a masked attention mechanism. This mechanism focuses only on partial positional information; specifically, it only computes the e_{ij} for the first-order neighboring nodes of node i and normalizes the attention coefficients of different nodes using the softmax function to obtain normalized coefficients α_{ij}. These normalized coefficients α_{ij} are then used to compare the attention coefficients among different nodes. The expression is as follows:

$$\alpha_{ij} = \text{softmax}(e_{ij}) = \frac{\exp(e_{ij})}{\sum_{k\in N_i}\exp(e_{ik})} \tag{15}$$

After obtaining the normalization coefficient α_{ij}, a nonlinear activation function is used to update the node's own features as output by linearly combining the features of adjacent nodes:

$$x'_i = \sigma\left(\sum_{j\in N_i} \alpha_{ij}\mathbf{W}x_j\right) \tag{16}$$

where N_i is the set of adjacency matrices for node i, and $\sigma(\cdot)$ is a nonlinear activation function. To enhance the stability of the self-attention learning process, the GAT employs a multi-head attention mechanism. This mechanism involves combining K separate attention mechanisms. After transforming the node features according to the aforementioned process, these mechanisms are concatenated to produce updated node output features. The resulting output features are as follows:

$$x'_i = \|_{k=1}^{K}\sigma\left(\sum_{j\in N_i} \alpha_{ij}^{k}\mathbf{W}^{k}x_j\right) \tag{17}$$

4.2. Deep Deterministic Policy Gradient (DDPG)

The ADNDR model for an active distribution network, which is dynamic in nature, poses a challenging high-dimensional, mixed-integer, nonlinear optimization problem that incorporates stochastic elements. Traditional optimization techniques, including mathematical algorithms and heuristic methods, face challenges in addressing this intricate problem, particularly regarding computational efficiency and precision. DRL interacts with its environment to continually refine behavioral strategies. Its ultimate goal is to secure the highest possible long-term average cumulative rewards, along with the respective optimal strategies. It also utilizes deep neural network approximation functions and policy functions to handle more complex state and action spaces, improving the applicability and performance of the algorithm.

The DRL algorithm has four advantages in solving the ADNDR problem: (1) The DRL algorithm has strong adaptability to uncertain factors. Traditional mathematical optimization algorithms have many shortcomings in dealing with uncertain factors, while DRL algorithms can learn and formulate optimal strategies in uncertain environments through interaction and feedback with the environment, thus adapting well to changes in uncertain factors. (2) The DRL algorithm does not require predicting load and renewable energy output. At present, most algorithms are based on distributed energy and load prediction data for ADNDR [15,16]. However, there exists a degree of discrepancy between the forecasted data and the real-world scenario. This discrepancy has the potential to trigger operational hazards, including voltage deviations beyond acceptable limits, excessive power loads, and heightened network losses. During the actual implementation of the reconfiguration strategy, these risks may manifest. The DRL algorithm possesses the capability to directly derive decisions from the present system state and the associated action value functions. Consequently, forecasting renewable energy and load output becomes unnecessary. (3) The DRL algorithm does not require distribution network parameters for model construction. For complex distribution network structures, parameter acquisition is often not directly possible, while the DRL algorithm can learn directly from the environment and find the optimal decision action through reward and punishment signal feedback from the environment, which belongs to a model-free algorithm. (4) The DRL algorithm takes into account the long-term returns associated with the operation of the distribution network. In the dynamic reconfiguration problem of active distribution networks, the state of the distribution network changes dynamically, including changes in load, topology, etc. The objective function is usually described as cumulative benefits over some time, as shown in Equations (1)–(4). DRL determines the most suitable reconfiguration strategy by taking into account long-term benefits and possessing a level of dynamic adaptability. This enables it to tackle the sequential decision-making optimization challenge in dynamic distribution networks efficiently, ultimately fulfilling the long-term operational requirements of these networks more effectively. In summary, the DRL algorithm is very suitable for solving the ADNDR problem.

Based on how intelligent agents make choices, reinforcement learning algorithms can be categorized into three distinct types. These include algorithms that are value-based, those that are policy-based, and those that integrate both value and policy considerations. Value-based algorithms are often able to better utilize data and learn more accurate estimates of value functions. Policy-based algorithms can often better explore the action space by directly parameterizing policies. The DDPG algorithm is seen as a combination of value-based and policy-based algorithms, which update parameters through approximation functions and deterministic strategies. Hence, it integrates the swift convergence characteristics of policy-based algorithms with the stability and reliable convergence traits of value-based approaches. The DDPG algorithm, utilizing the actor–critic architecture, is capable of efficiently leveraging data throughout the learning phase. It achieves a balance between exploring new possibilities and utilizing known information. The structure diagram of DDPG is demonstrated in Figure 2. In Figure 2, it can be found that the DDPG model consists of two neural networks: one actor network θ^{μ} is used to decide the action at the current moment, and the other critic network θ^{Q} is used to estimate the action value function and the quality of the current state value. DDPG stores all transition states ($\boldsymbol{s}_t$, $\boldsymbol{a}_t$, r_t, $\boldsymbol{s}_{t+1}$) experienced in the experience replay pool D, where $\boldsymbol{s}_t$ represents the current state, $\boldsymbol{a}_t$ represents the current action, r_t represents the reward function value, and $\boldsymbol{s}_{t+1}$ represents the next state.

The actor policy network in DDPG is divided into the current network θ^{μ} and the target network $\theta^{\mu'}$. The current network θ^{μ} selects the optimal action based on the current state $\boldsymbol{s}_t$ provided by the environment, thereby generating the next state $\boldsymbol{s}_{t+1}$, obtaining a reward r_t, and updating the current network parameters θ^{μ} based on the Q value calculated by the critic network θ^{Q}. The target network selects the optimal next action $\boldsymbol{a}_{t+1}$ based on the next state $\boldsymbol{s}_{t+1}$ sampled from the experience replay pool. The target network parameters $\theta^{\mu'}$ are updated at fixed intervals based on the current network parameters θ^{μ}.

For the actor architecture, traditional implementations employ MLPs to process state features. In contrast, our GAT-DDPG framework introduces Graph Attention Layers to model state representations as graph-structured data. Specifically, the GAT dynamically assigns attention weights to neighboring nodes through learnable coefficients, enabling the actor to focus on critical interactions (e.g., agent dependencies in multi-agent systems) while filtering irrelevant connections. This replaces MLP's fixed-weight aggregation with adaptive relational reasoning, significantly enhancing action generation in environments with implicit or evolving dependencies.

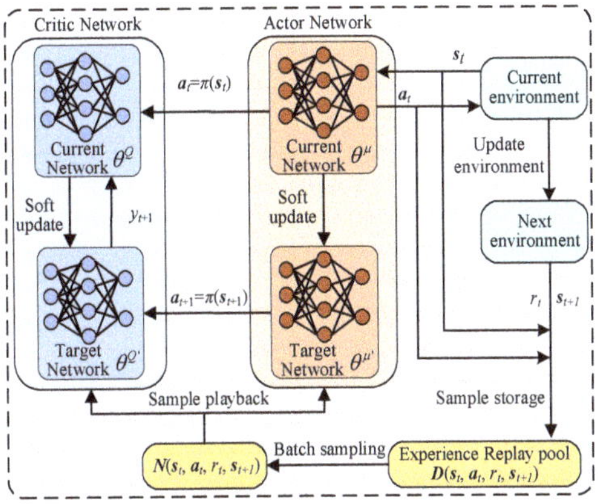

Figure 2. DDPG structure diagram.

The critic network in DDPG is divided into the current network θ^Q and the target network $\theta^{\mu'}$. The current network θ^Q calculates the current Q value $Q(s_t, a_t, \theta^Q)$ based on the action a_t selected by the actor current network θ^μ and the current state s_t, which is used to update the actor's current network parameters and the commentator's current network parameters θ^Q. The target network is responsible for calculating $Q'\left(s_{t+1}, a_{t+1}, \theta^{Q'}\right)$ in the target Q value y_i, and the expression for the target Q value y_i is as follows:

$$y_i = r_t + \gamma Q'\left(s_{t+1}, a_{t+1}, \theta^{Q'}\right) \tag{18}$$

where y_i is the target Q value; γ is the attenuation factor; and $Q'\left(s_{t+1}, a_{t+1}, \theta^{Q'}\right)$ is the Q value calculated by the target critic network $\theta^{Q'}$ based on the next state s_{t+1} and the optimal action a_{t+1}. The network parameters $\theta^{\mu'}$ are updated at fixed intervals based on the current network θ^μ.

Unlike standard DDPG, where the critic concatenates state–action vectors as MLP inputs, our GAT-based critic constructs a hybrid graph where nodes encode state features and edges integrate action attributes. GAT layers propagate features across this graph, computing attention-based Q-values that capture both local action impacts and global interactions (e.g., long-term dependencies in continuous control tasks). This design ensures nuanced modeling of state–action interdependencies, particularly in partially observable environments.

The goal of the actor network in DDPG is to obtain a larger Q value as much as possible, and the smaller the feedback Q value obtained, the greater the loss. Therefore, as long as a negative sign is taken on the Q value returned by the commentator's current network θ^Q, it is called the loss function, and its expression is as follows:

$$J(\theta^\mu) = -\frac{1}{m}\sum_{i=1}^{m} Q\left(s_t, a_t, \theta^Q\right) \tag{19}$$

where $J(\theta^\mu)$ is the loss function value of the actor current network θ^μ; m is the number of samples with batch gradient descent; and $Q(s_t, a_t, \theta^Q)$ is the Q value calculated by the critic network θ^Q based on the current state s_t, current action a_t, and current network parameters θ^Q. The actor current network θ^μ updates the network parameters using a backpropagation algorithm based on the loss function.

The commentator in DDPG's current network θ^Q loss function is a mean square error, which is expressed as follows:

$$J\left(\theta^Q\right) = \frac{1}{m}\sum_{i=1}^{m}\left(y_i - Q\left(s_t, a_t, \theta^Q\right)\right)^2 \tag{20}$$

where $J(\theta^Q)$ is the loss function value of the commentator's current network θ^Q; m is the number of samples with batch gradient descent; y_i is the target Q value; and $Q(s_t, a_t, \theta^Q)$ is the Q value calculated by the critic's network θ^Q based on the current state s_t, current action a_t, and current network parameters θ^Q. The critic current network θ^Q uses a backpropagation algorithm to update network parameters based on this loss function.

The soft-update technique is utilized for modifying the settings of the actor target network, as detailed below:

$$\theta^{\mu'} \leftarrow \tau\theta^{\mu} + (1 - \tau)\theta^{\mu'} \tag{21}$$

where τ is the update coefficient, and in this paper, $\tau = 0.01$ is used.

5. An ADNDR Method Based on GATDDPG

The complexity of the distribution network leads to a situation where the interconnections among nodes play a crucial role in determining the power system's stability. Furthermore, these connections have a profound impact on its reliability. The GAT excels in handling graph-structured data with complex topological structures, while DDPG is an efficient reinforcement learning algorithm that can learn efficient and stable reconfiguration strategies by exploring in the distribution network environment. Therefore, the research presented in this paper introduces a novel approach to ADNDR, utilizing the GATDDPG framework. By introducing the GAT to process graph structure information in the environment, the proposed algorithm can more accurately understand the dynamics and relationships of the environment. Meanwhile, by leveraging DDPG for optimizing and rearranging the decision-making procedure, we can guarantee the agent acquires strategies that are both stable and effective. This section constructs a reinforcement learning model for the ADNDR strategy based on GATDDPG, as shown in Figure 3.

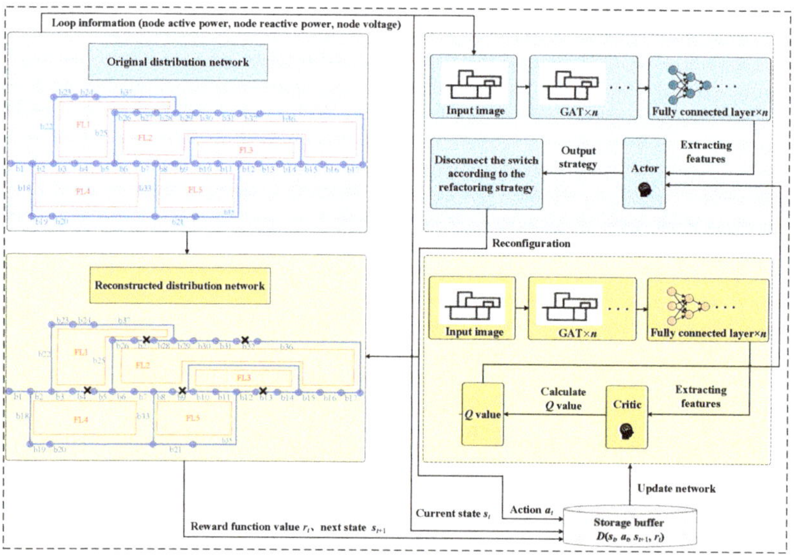

Figure 3. Structure diagram of the ADNDR strategy based on GCNDDPG.

The Markov decision process (MDP) is aimed at making decisions through a random program with Markov properties [17]. The MDP provides a solid theoretical foundation for reinforcement learning, allowing us to make optimal decisions while considering the current state and possible future changes in the distribution network. By modeling the distribution network reconfiguration problem as the MDP, we can effectively utilize the GATDDPG algorithm to search and learn the optimal distribution network reconfiguration strategy. The following describes the MDP model for ADNDR, including the state s_t, action a_t, and reward function r_t.

5.1. State

The state $s_t = (\boldsymbol{G}_t, \boldsymbol{X}_t)$ includes the topology diagram $\boldsymbol{G}_t$ and distribution network node characteristic (voltage, power, etc.) information $\boldsymbol{X}_t \in R^{N\times C}$ at time t. C is the number of features of the node, where $C = 3$; N is the number of distribution network nodes, where $N = 33$.

(1) The distribution network topology diagram $\boldsymbol{G}_t$ represents the connection relationship of nodes in the distribution network at time t.
(2) The node feature matrix $\boldsymbol{X}_t$ represents the node feature information of the distribution network at time t, and each row represents a node feature vector. Its expression is as follows:

$$\boldsymbol{X}_t = \{\boldsymbol{P}_t, \boldsymbol{Q}_t, \boldsymbol{V}_t\} \tag{22}$$

where $\boldsymbol{P}_t$ and $\boldsymbol{Q}_t$ are the sets of active and reactive power of each node at time t, respectively, and $\boldsymbol{V}_t$ is the set of voltages at each node at time t. In this study, $\boldsymbol{X}_t$ was calculated using the Newton–Raphson method, a robust iterative algorithm widely adopted in power system analysis for solving nonlinear power flow equations. By iteratively linearizing nonlinear equations using Taylor series expansion (ignoring higher-order terms), it converts the problem into solving a series of linear matrix equations (e.g., Jacobian matrix updates) until convergence.

5.2. Action

The action generally refers to the strategies or behaviors that intelligent agents can adopt. On this basis, the intelligent agent selects a certain action to execute in the action space based on the state s_t, that is:

$$a_t = \boldsymbol{k},\ \boldsymbol{k} \in \boldsymbol{S}_{\text{cheme}} \tag{23}$$

Equation (23) indicates that the agent selects an action $\boldsymbol{a}_t$ from the action space S_{cheme}. In our study, the action is chosen based on the encoding technique outlined in Section 2. Specifically, there are 732 viable options available, meaning we select one of these 732 solutions for execution. For example, in the IEEE 33-bus power system shown in Figure 1, if a selected five-dimensional action vector is {0, 1, 0, 2, 3}, as shown in Table 3, the action scheme represents the disconnection of branch 3, branch 8, branch 9, branch 18, and branch 35 and the closure of other remaining branches in the distribution network.

5.3. Reward Function

The reward function constitutes a fundamental element in reinforcement learning algorithms. It outlines the agent's learning goal, namely, achieving the highest possible accumulation of rewards. This paper transforms the objective function in the mathematical model of distribution network reconfiguration in Section 2 into a reward function to reduce network loss and voltage offset, which is helpful for the training of intelligent agents. At the same time, it is necessary to ensure that the ADNDR solution of the agent can satisfy all constraint conditions; otherwise, the results of distribution network reconfiguration will be meaningless. Therefore, this paper establishes the following reward function:

$$r_t = \begin{cases} -f_{\text{ADNDR}}, & f_{\text{success}} = 1 \\ -3, & f_{\text{success}} = 0 \end{cases} \tag{24}$$

where r_t is the reward value for executing action $\boldsymbol{a}_t$ based on state $\boldsymbol{s}_t$ at time t; when $f_{\text{success}} = 1$, it indicates that the proposed reconfiguration scheme fully complies with all the established constraint conditions in the ADNDR mathematical model, specifically encompassing the contents described in Equations (7) to (13). At this point, the reward value will be calculated normally according to the established rules. Conversely, if $f_{\text{success}} = 0$, it implies that the reconfiguration scheme fails to fully meet the constraint conditions in the mathematical model. In this unfavorable scenario, in order to guide the agent to avoid such schemes in future decisions, we will impose a significant penalty on the agent, which is set to -3 in this study.

5.4. Solving Process

A GATDDPG algorithm was developed for addressing the challenge of ADNDR. This allows us to attain the best approach for rearranging the network configuration. The procedure for dynamically reconfiguring an active distribution network while considering graph structure information is depicted in Figure 4. Below are the detailed steps involved in this process:

Step 1: Initialize the current network parameters θ^{μ} of actors, actor target network $\theta^{\mu'}$, commentator current network parameters θ^{Q}, commentator target network parameters $\theta^{Q'}$, and iteration number e in GATDDPG and determine the maximum iteration number E and maximum random exploration number M.
Step 2: Adopt a coding method based on basic circuits to reduce the action space of ADNDR.
Step 3: Initialize the time identifier t.
Step 4: Initialize the distribution network and obtain the voltage, active power, and reactive power of each node according to Equation (22), as well as the distribution network structure diagram $\boldsymbol{G}_t$. Construct the initial state space $\boldsymbol{s}_t = (\boldsymbol{G}_t, \boldsymbol{X}_t)$.
Step 5: If the number of iterations is less than the maximum number of random explorations M, the actor adopts a random policy. If the number of iterations is greater than the maximum number of random explorations M, the actor network θ^{μ} selects the optimal action $\boldsymbol{a}_t$ based on the current distribution network state $\boldsymbol{s}_t$.
Step 6: According to action $\boldsymbol{a}_t$, disconnect the corresponding switch in the distribution network and reconfigure the network.
Step 7: Perform power flow calculation on the reconfigured distribution network, obtain the reconfigured state space $\boldsymbol{s}_{t+1}$, and calculate the reward function value r_t according to Equation (24).
Step 8: Store the pre-refactoring state $\boldsymbol{s}_t$, refactoring action $\boldsymbol{a}_t$, corresponding reward function value r_t, and the refactored state $\boldsymbol{s}_{t+1}$ in the experience replay pool D.
Step 9: If the number of iterations is greater than the number of random explorations M, proceed to step 10. If the number of iterations is less than the random exploration number M and the time identifier is less than 24, update the time and return to step 5. If the iteration count is less than the random exploration count M and the time identifier is greater than or equal to 24, update the iteration count and return to step 3.
Step 10: Batch sample m samples $\{\boldsymbol{s}_t, \boldsymbol{a}_t, r_t, \boldsymbol{s}_{t+1}\}$ from the experience replay pool.
Step 11: Use the commentator to calculate the Q value between the current network θ^{Q} and the target network $\theta^{Q'}$ based on Equation (18) and the loss function value through

Equations (19) and (20) based on the Q value. Update the network parameters using the backpropagation algorithm. When the time identifier reaches 24, execute step 12. Otherwise, update the time and execute step 5.
Step 12: Terminate the training process once the predefined maximum iteration limit is attained. If not, increment the iteration count and advance to the next step, which is step 3.

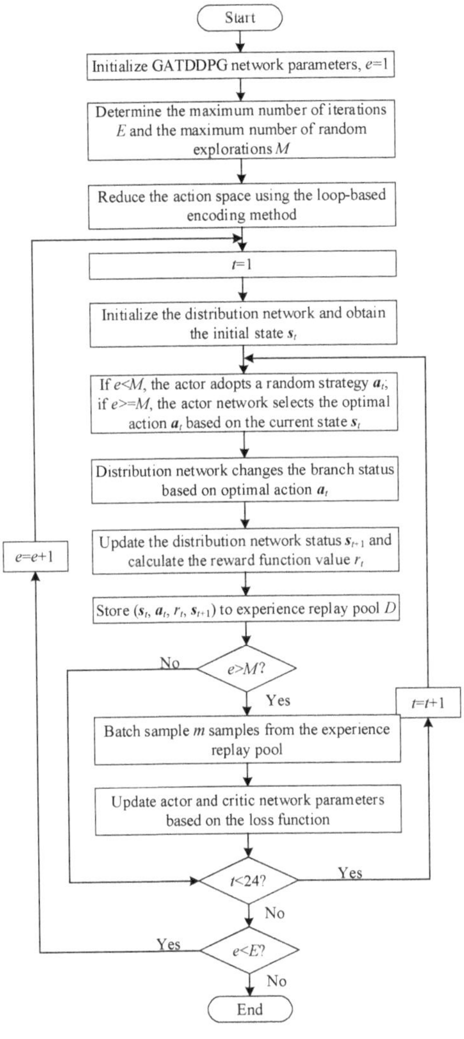

Figure 4. Flowchart of the dynamic reconfiguration algorithm for active distribution networks.

6. Case Study

6.1. Simulation Environment Settings

This paper presents an enhanced version of the IEEE 33-bus power system, specifically tailored for case studies, which is depicted in Figure 5. The IEEE 33-bus power system consists of 1 substation and 37 branches, with a voltage range of 0.9~1.1 during normal operation. In this paper, $H_{\mathrm{SWI,max}}$ and $H_{\mathrm{SWI,b,max}}$ are set to 15 and 3, respectively [18]. The basic branch and load information in the IEEE's 33 nodes can be found in reference [19]. The permissible capacity limits of the branches strictly adhere to the values documented in [14] (Table 1) to ensure operational safety constraints. The wind turbines with rated powers of 600 kW, 1100 kW, and 1000 kW are installed at nodes 10, 18, and 21 of the IEEE 33-bus power system, respectively. Photovoltaic generators with rated powers of 600 kW, 1100 kW, and 1000 kW are installed at nodes 7, 15, and 26 of the IEEE 33-bus power system, respectively. And the power factors are set as 0.9. The 24 h photovoltaic generation, wind turbine, and load power curves depicted in Figure 6 are long-term averaged profiles derived from multi-year historical data (2015–2024) via the Xihe Energy Big Data Platform [20]. The photovoltaic curves integrate hourly solar irradiance and temperature data calibrated with the DISC model to account for spectral variations and panel efficiency degradation, while the wind power curves incorporate aerodynamic corrections and air density adjustments based on altitude and temperature. The load profiles reflect regional demand patterns synthesized from industrial, commercial, and residential consumption trends, excluding extreme weather scenarios. During training, stochastic fluctuations (±5%) are superimposed on these averaged curves to simulate real-time uncertainties while maintaining physical feasibility.

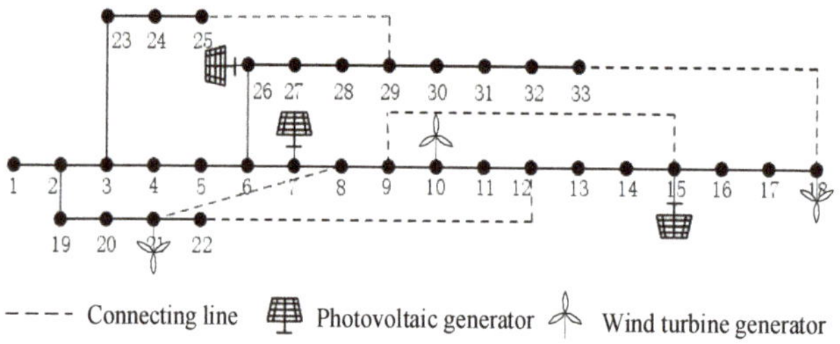

Figure 5. Schematic diagram of an improved IEEE 33-bus system.

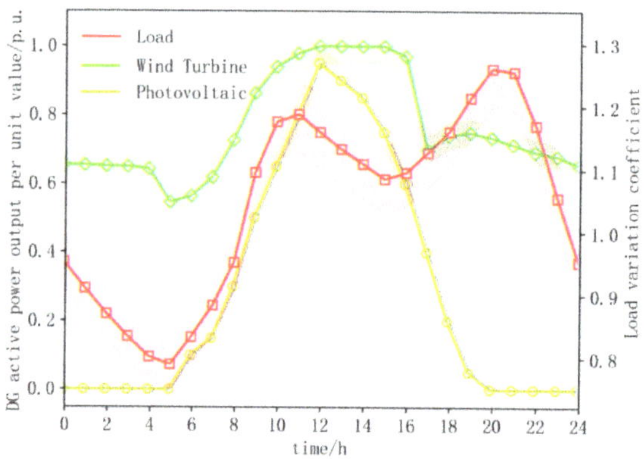

Figure 6. The 24 h DG active power and load curves.

6.2. Simulation Parameter Settings

The parameters and neural network model of GATDDPG are shown in Tables 4 and 5. It can be found in Table 5 that the actor network θ^{μ} consists of an input layer, three hidden layers, and one output layer. In the input layer, the state includes active power, reactive power, and the voltage of all nodes, whose dimension is 33 × 3. Thus, the input layer of the actor network is a 33 × 3 graph feature matrix, which is activated by two GAT layers and ReLU functions and then by two FC layers to finally output the policy function of the current state.

Table 4. Parameters of GATDDPG algorithms.

Parameter	Value
Learning rate	5×10^{-5}
Discount factor	0.9
Training batch size	64
Storage capacity	10,000
Iterations	1000
Random exploration times	300

Table 5. The structure and activation function of the neural network in GATDDPG.

Network Layer	Actor		Critic	
	Neural Network Structure	Activation Function	Neural Network Structure	Activation Function
Input layer	33×3	/	33×3	/
GAT1	$3 \rightarrow 32$	ReLU	$3 \rightarrow 32$	ReLU
GAT2	$32 \rightarrow 16$	ReLU	$32 \rightarrow 16$	ReLU
FC1	$33 \times 16 \rightarrow 64$	ReLU	$33 \times 16 + n_{\text{action}} \rightarrow 64$	ReLU
FC2 (output layer)	$64 \rightarrow n_{\text{action}}$	softmax	$64 \rightarrow 1$	/

The critic network θ^Q in this paper consists of one input layer, three hidden layers, and one output layer. The input layer is also a 33×3 graph feature matrix, which is activated by two layers of GAT and ReLU functions and then by two layers of fully connected (FC) layers to finally output the action value function of the current state.

6.3. Hyperparameter Measurement

The determination of the learning rate along with the discount factor has a substantial influence on the efficacy of agent training. Consequently, it is necessary to conduct a thorough examination and contrast of these parameters to ascertain the most suitable hyperparameter setup. The reward function values of GATDDPG under different learning rates and discount factors are presented in Table 6. Meanwhile, the optimization results of GATDDPG under different learning rates and discount factors are shown in Figures 7 and 8. Observing Figure 7, it is evident that when the learning rate is set to 5×10^{-6}, the optimization effectiveness of the GATDDPG algorithm is not ideal, attributed to the excessively low learning rate, which results in a small step size during parameter updates, making it difficult to effectively capture gradient information. However, when the learning rate is set to 5×10^{-5} and 5×10^{-6}, GATDDPG exhibits good optimization performance. Among them, it is clear from Table 6 that the reward function value at a learning rate of 5×10^{-5} is higher than that at a learning rate of 5×10^{-6}. Therefore, the final learning rate for GATDDPG is determined to be 5×10^{-5}. Observing Figure 8, when the discount factor is set to 0.95, the GAT requires a longer training period to achieve convergence of the reward function value, due to the excessively high discount factor, which makes the algorithm overly focused on long-term rewards while neglecting near-term feedback. In contrast, the GATDDPG algorithms with discount factors of both 0.9 and 0.85 exhibit good optimization performance. Among them, the network with a discount factor of 0.9 converges slightly faster than that with 0.85. And it is clear from Table 6 that the reward function value at a discount factor of 0.9 is higher than that at a discount factor of 0.85. However, it should be noted that although the discount factor of 0.95 performs poorly in the early stages of training, its long-term performance may be better, requiring a trade-off based on specific application scenarios. However, based on the current data and requirements, the final discount factor is determined to be 0.9. In summary, the learning rate of the GATDDPG network used in this paper is finally set to 5×10^{-5}, and the discount factor is set to 0.9.

Table 6. Comparison of evaluation indicators for different hyperparameters.

Algorithms	Learning Rate	Discount Factor	Reward
GATDDPG	5×10^{-5}	0.85	−3.6544
	5×10^{-5}	0.9	−2.3621
	5×10^{-5}	0.95	−3.1141
	5×10^{-6}	0.95	−3.5194
	5×10^{-4}	0.95	−2.5413

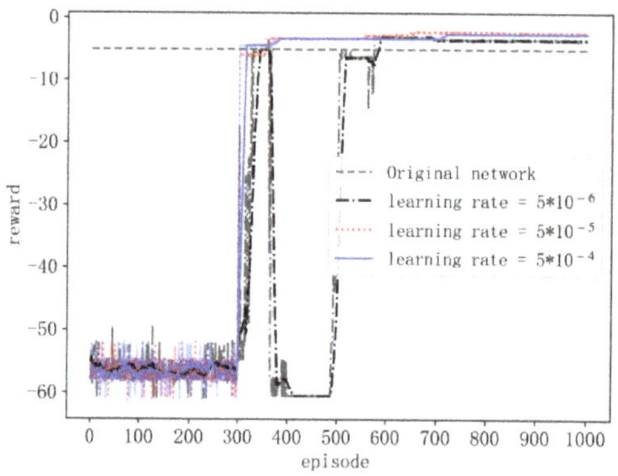

Figure 7. Optimization results under different learning rates.

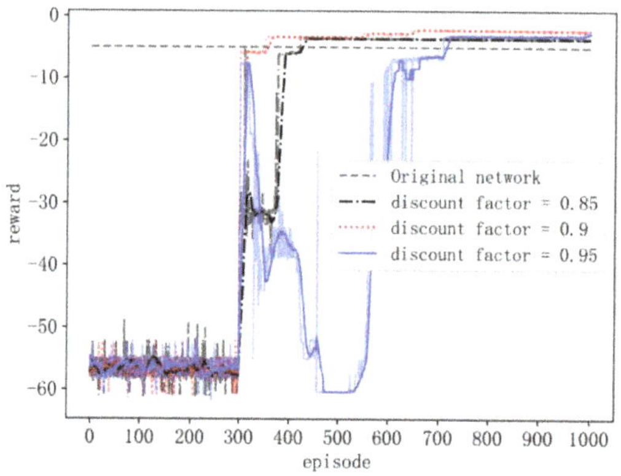

Figure 8. Optimization results under different discount factors.

6.4. Comparative Analysis of Network Loss and Voltage Under Different Algorithms

To showcase the efficacy and advantage of our GATDDPG-inspired optimization method for ADNDR strategies, the present section evaluates its performance by contrasting it with four alternative optimization algorithms, all in the context of tackling ADNDR strategy optimization challenges. Specifically, these algorithms include the MISOCP algorithm, the standard DDPG algorithm, the GATDDQN algorithm, and the GCNDDPG algorithm.

Firstly, we aim to showcase the viability of employing deep neural networks for approximating the action value function. To this end, Figure 9 presents the loss function values obtained using various algorithms. Since the training of the neural network only starts from the 300th episode, the range of episode values is from 300 to 1000.

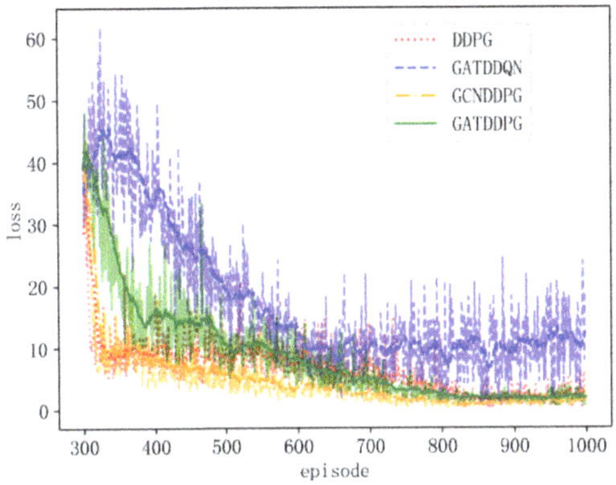

Figure 9. Value of the loss function under different algorithms.

From Figure 9, it can be seen that the loss functions of the neural networks of each DRL algorithm can effectively converge, and their training process is fast and stable. Therefore, deep neural networks can accurately predict the action value function in various DRL algorithms.

Furthermore, to rigorously evaluate the effectiveness of the GATDDPG-based approach for ADNDR, we undertook a comparative study. This involved assessing the performance of multiple algorithms in tackling ADNDR challenges. Specifically, the reward function values under different algorithms are shown in Figure 10, while a comparison of the evaluation indicators for different algorithms is presented in Table 7. To quantitatively validate the effectiveness of the GATDDPG-based approach (ADNDR), a comparative analysis of various algorithms' performance on ADNDR was conducted, as illustrated in Figure 10 and summarized in Table 7.

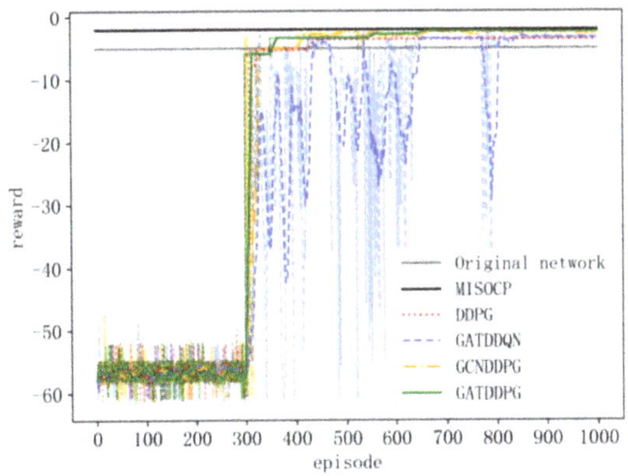

Figure 10. Reward function values under different algorithms.

Based on the various algorithms, the ADNDR strategy is capable of enhancing the performance metrics of the distribution network. Specifically, it improves network loss and reduces voltage deviation, as illustrated in Figure 10. The ADNDR strategy utilizing GATDDQN exhibits relatively sluggish reward convergence. In contrast, the ADNDR strategy leveraging GCNDDPG achieves faster convergence, albeit with a reward function that slightly underperforms compared to the ADNDR strategy grounded in GATDDPG presented herein.

Table 7. Comparison of evaluation indicators for different algorithms.

Algorithms	Reward	Training Time/h	Decision Time/s
Original network	−5.0478	/	/
MISOCP	−2.2263	/	34.94
DDPG	−3.6378	2.08	0.142
GATDDQN	−3.3293	1.33	0.130
GCNDDPG	−2.7082	1.82	0.133
GATDDPG (The proposed method)	−2.3615	0.78	0.111

From Table 7, we can conclude the following:

(1) In terms of reward values, the reward function of GATDDPG exhibits higher numerical values compared to GATDDQN, indicating that the DDPG algorithm has advantages in optimization performance over the DDQN algorithm. Furthermore, both GATDDPG and GCNDDPG surpass the traditional DDPG in terms of reward values, demonstrating the necessity and effectiveness of considering graph structure information when solving the ADNDR problem. Specifically, the reward value of GATDDPG reaches −2.3615, achieving a 12.8% improvement compared to the −2.7082 of GCNDDPG, which strongly proves that using the GAT to fit the action-value function is superior to the GCN.

(2) In terms of computing speed, the training time of GATDDPG is 0.68 h, which is reduced by 48.8%, 63.6%, and 12.8% compared to the training time of DDPG, GATDDQN, and GATDDPG, respectively. The decision times of GATDDPG, GCNDDPG, GATDDQN, DDPG, and MISOCP are 0.111 s, 0.133 s, 0.130 s, 0.142 s, and 34.94 s, respectively. Compared with GCNDDPG, GATDDQN, DDPG, and MISOCP, the decision time of GCNDDPG is reduced by 16.5%, 14.6%, 21.8%, and 99.68%, respectively.

(3) In terms of constraint conditions, based on the definition in Equation (24), this study assigns a penalty factor of −3 to reconfiguration schemes that fail to meet the constraints within each time step. During the random exploration phase, the agent adopts a strategy of randomly selecting actions, which may result in the selected actions not fully complying with all constraints, thereby causing a significant decrease in the cumulative reward function value, specifically to around −60. However, upon entering the optimization phase, except for the GATDDQN algorithm, the 24 h cumulative reward function values of the other algorithms stabilize near −3. This phenomenon indicates that the reward function value at each time step exceeds −3, implying that the penalty term is not activated. Therefore, we can reasonably infer that the GATDDPG-based ADNDR strategy can effectively satisfy all the constraint conditions listed in the mathematical model (i.e., Equations (7)–(13)).

To validate the effectiveness of the proposed method, this study systematically compared the performance differences of various algorithms across key metrics, including daily average network loss, voltage deviation, node voltage distribution at 12:00, and dynamic network loss, with the experimental results illustrated in Figures 11–13. Data analysis reveals the following: Under the benchmark scenario, the distribution network exhibited daily average network loss and voltage deviation of 135.0650 kW and 0.8073 p.u., respectively. The DDPG algorithm reduced these metrics to 77.3583 kW and 0.7451 p.u. (reductions of 42.73% and 7.70%), while GATDDQN and GCNDDPG further optimized them to 65.2251 kW/0.4697 p.u. (reductions of 51.71%/41.82%) and 67.3276 kW/0.5862 p.u. (reductions of 50.15%/27.39%), respectively. Although all aforementioned DRL algorithms improved network performance, their optimization margins differed significantly.

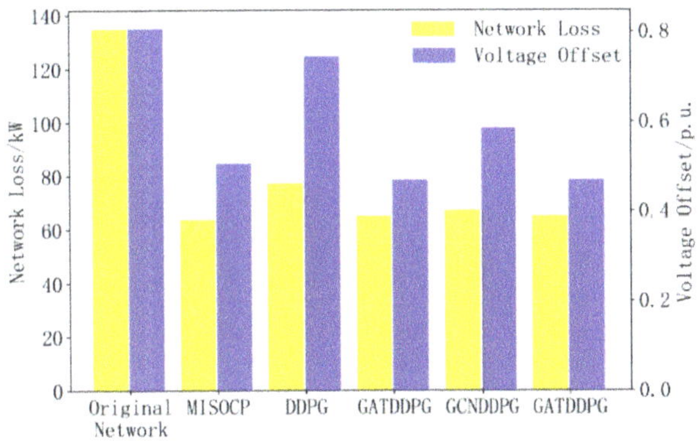

Figure 11. Daily average network loss and voltage deviation for different algorithms.

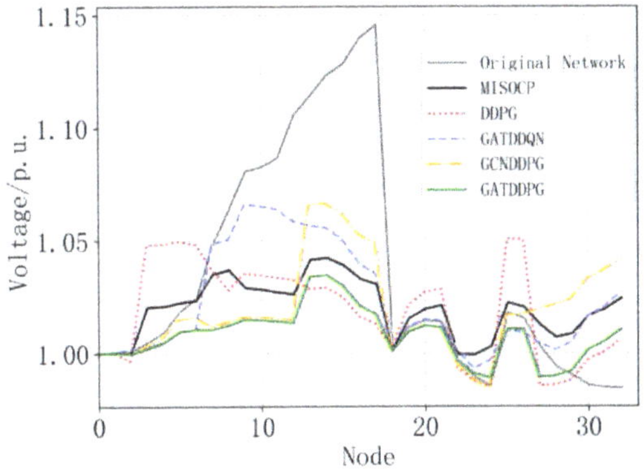

Figure 12. Voltage at each node at 12:00 for different algorithms.

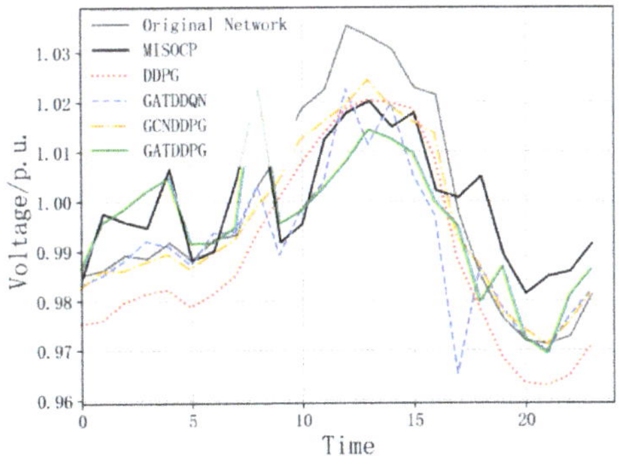

Figure 13. Average node voltage of different algorithms at different times.

Notably, the GATDDPG-based ADNDR strategy achieved the highest reductions in both metrics (51.71% network loss and 41.82% voltage deviation), outperforming the conventional DRL methods. In contrast, the MISOCP mathematical optimization method further reduced the metrics to 63.9830 kW/0.5071 p.u. (reductions of 52.63%/37.18%), but at the cost of significantly degraded computational efficiency—its solving time grew exponentially with network scale, rendering it impractical for real-time optimization. The proposed method maintained competitive optimization performance (with only 0.92% and 4.64% gaps in network loss and voltage deviation compared to MISOCP) while reducing the computation time by two orders of magnitude. Furthermore, as illustrated in Figures 12 and 13, in terms of node voltage stability, the method based on GATDDPG demonstrated superior performance compared to MISOCP. Additionally, Figure 14 shows that the network loss trajectory optimized by the reconstruction strategy based on GATD-DPG was extremely close to the results obtained by MISOCP. These observations robustly validate the exceptional global optimization capability of the proposed method.

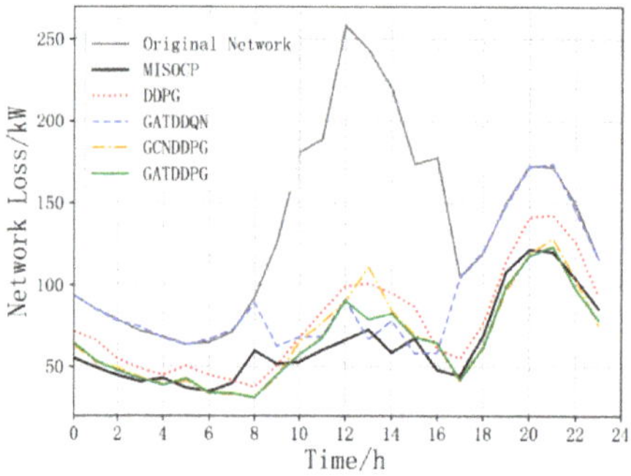

Figure 14. Network loss at various time points for different algorithms.

These experiments demonstrate that the GATDDPG-based ADNDR strategy effectively captures implicit topological correlations through graph attention mechanisms, achieving near-optimal control performance without requiring precise parameter identification. This provides a highly efficient and robust solution for real-time reconfiguration in large-scale active distribution networks.

As shown in Table 8, a more in-depth analysis of the dynamic reconfiguration strategy was conducted. The topology reconfiguration strategy generated by ADNDR based on GATDDPG exhibits significant temporal variability in the switching combinations across different time intervals. Specifically, the switching configurations during the 10:00 to 15:00 period differ significantly from those in other time periods. This difference can be attributed to the significant fluctuations in PV output during the peak irradiation period (i.e., 10:00–15:00), as visually demonstrated by the PV generation curve in Figure 6.

Table 8. Dynamic network reconfiguration strategy.

Time	Reconfiguration Switch
0:00–9:00	[37, 34, 13, 7, 8]
10:00–15:00	[27, 8, 13, 33, 9]
16:00–23:00	[37, 34, 13, 7, 8]

Additionally, Table 9 illustrates the action switches involved in the reconfiguration strategy along with their corresponding action counts. In Figure 9, it can be observed that the action count for each individual switch is less than the maximum allowed action count for that switch (3 times/day), and the total action count for all switches is also less than the maximum total action count (15 times/day). This indicates that the ADNDR strategy based on GATDDPG can optimize network loss and voltage deviation in the distribution network while satisfying the constraints on switch action counts.

Furthermore, Figure 13 reveals that as the PV output increases, there is a sharp rise in distribution network voltage. This phenomenon prompts the GATDDPG algorithm to dynamically adjust and reconfigure its strategy by prioritizing voltage-sensitive nodes. This adaptive feature of the algorithm fully demonstrates its sensitivity to real-time operating conditions, thereby effectively ensuring voltage stability in environments with high renewable energy penetration rates.

Table 9. The number of switch actions corresponding to the dynamic network reconfiguration strategy.

Action Switch	Number of Switch Operations	Sum of Switch Operation Counts
b7	2	12
b9	2	
b27	2	
b33	2	
b34	2	
b37	2	

6.5. Analysis of Adaptability to Uncertainty Factors

To further validate the superiority of the proposed ADNDR based on GATDDPG in dealing with uncertainties, we conducted a comparative analysis of the following four optimization strategies:

Strategy 1: Without considering the uncertainties in load and DG output, the MISOCP algorithm was used to solve the ADNDR strategy.
Strategy 2: Fully considering the uncertainties in load and DG output, the MISOCP algorithm was also employed to solve the ADNDR strategy.
Strategy 3: Ignoring the uncertainties in load and DG output, the GATDDPG algorithm was utilized to solve the ADNDR strategy.
Strategy 4: Taking into full account the uncertainties in load and DG output, the GATDDPG algorithm was applied to implement the solution for the ADNDR strategy.

The optimization results for each strategy are detailed in Table 10. According to the data in Table 10, we can observe that in environments containing uncertainties, the reward function value of the MISOCP algorithm decreased from −2.0628 to −2.2263, representing a decline of 7.9%. In contrast, the optimization performance of the GATDDPG algorithm was not compromised by the presence of uncertainties in the environment. Instead, its reward function value increased, which strongly demonstrates the robust adaptability of the GATDDPG algorithm in dealing with uncertainties.

Table 10. Reward function values for different optimization methods.

Strategy	Reward
Strategy 1	−2.0628
Strategy 2	−2.2263
Strategy 3	−2.4686
Strategy 4	−2.3615

6.6. Impact of Operational Constraints on Network Reconfiguration Decision Consistency

To delve deeper into the specific impacts of constraints on reconfiguration strategies, this subsection employs the GATDDPG algorithm to solve the following four mathematical models:

Model 1: Model 1 sets Equation (4) as the optimization objective, with Equations (7)–(11) serving as constraints. This model does not incorporate the constraints of switch operation frequency and branch transmission capacity.
Model 2: Model 2 sets Equation (4) as the optimization objective and introduces Equations (7)–(11) and (13) as constraints. This model considers the constraint of branch transmission capacity but does not consider the constraint of switch operation frequency.

Model 3: Model 3 sets Equation (4) as the optimization objective and specifies Equations (7)–(12) as constraints. This model considers the constraint of switch operation frequency but does not consider the constraint of branch transmission capacity.
Model 4: Model 4 sets Equation (4) as the optimization objective, with Equations (7)–(13) serving as constraints. This model simultaneously considers the constraints of switch operation frequency and branch transmission capacity.

The solutions to these four mathematical models are detailed in Table 11. Analysis of the data in Table 11 reveals that after considering the constraint of switch operation frequency, the reward function values of the reconfiguration strategies exhibit a downward trend, indicating that the constraint of the switch operation frequency has a significant impact on the distribution network reconfiguration strategies. However, when the number of switch operations in the distribution network exceeds a certain threshold, necessary maintenance or replacement operations are required. Therefore, although the reward function values derived from Models 1 and 2 are higher than those from Models 3 and 4, their reconfiguration strategies only possess theoretical exploration value and lack practical application significance due to their neglect of the switch operation frequency constraint. Furthermore, by comparing the solutions of Models 1 and 2, as well as Models 3 and 4, it can be observed that in the distribution network environment set in this paper, whether the constraint of branch transmission capacity is considered does not have a significant impact on reconfiguration strategies. However, this does not mean that considering the constraint of branch transmission capacity is unimportant, as it ensures the stable operation of the power grid within a safe range and effectively avoids potential risks such as overloading. In summary, the optimization method discussed in this paper can effectively optimize distribution network reconfiguration strategies under the premise of satisfying the constraints of the switch operation frequency and branch transmission capacity, providing a decision support tool that balances safety and economy for active distribution networks with a high proportion of renewable energy.

Table 11. Reward function values of different mathematical models.

Model	Reward
Model 1	−2.2043
Model 2	−2.2043
Model 3	−2.3615
Model 4	−2.3615

7. Conclusions

This research presents a dynamic adjustment approach for active distribution networks, utilizing the GATDDPG algorithm as its foundation. The method focuses on reconfiguring the network efficiently. This algorithm effectively combines the data processing ability of a GAT graph with the decision-making ability of DDPG. The main conclusions are as follows:

(1) Compared with the traditional encoding methods, the loop-based encoding method used in this paper greatly reduces the number of feasible solutions by 95.53%, indicating that this encoding method has higher solving efficiency.
(2) Compared with the DDPG-based ADNDR strategy, the proposed GATDDPG-based ADNDR strategy reduces network loss and voltage offset by 8.98% and 19.69%, respectively. This demonstrates that considering graph structure information is very effective in solving ADNDR.

(3) Compared with the ADNDR strategy based on GCNDDPN, the proposed ADNDR strategy based on GATDDPG reduces network loss and voltage offset by 1.56% and 14.43%, respectively. This indicates that, compared to using the GCN, using the GAT as a function fitting tool for graph reinforcement learning algorithms in solving ADNDR problems is more effective.

(4) Compared with the ADNDR strategy based on GCNDDPG, GATDDQN, DDPG, and MISOCP, the ADNDR strategy based on GATDDPG reduces the decision time by 16.5%, 14.6%, 21.8%, and 99.68%, respectively, indicating that this method has higher reconfiguration efficiency.

In summary, the proposed ADNDR strategy based on GATDDPG can better adapt to distribution networks containing DGs. At present, most research focuses on the economic or reliability optimization of distribution networks. In the future, our work will focus on the synergistic optimization of economics and reliability in distribution networks and solving more complex ADNDR problems.

Author Contributions: C.G. took part in conceptualizing this study alongside C.J. They both developed the methodology, with C.G. taking the lead. C.G., C.J. and C.L. collaborated on the validation. Data curation responsibilities fell to C.L. The initial draft of this manuscript was prepared by C.G. and C.J., who also reviewed and edited it subsequently. Visualization tasks were handled by C.L. and C.G. oversaw the entire project. Both C.G. and C.J. managed the project administration. Funding acquisition was a joint effort between C.L. and C.G. All authors have read and agreed to the published version of the manuscript.

Funding: This research was funded by the Natural Science Foundation of the Fujian Province, grant number 2022J05125, the Natural Science Foundation of the Fujian Province, grant number 2021J05134, the National Natural Science Foundation of China, grant number 52377087, and the National Natural Science Foundation of China, grant number 72401069.

Data Availability Statement: The original contributions presented in this study are included in this paper; further inquiries can be directed to the corresponding authors.

Conflicts of Interest: The authors declare no conflicts of interest.

References

1. Zhang, Y.; Zeng, B.; Zhou, Y.; Xu, H.; Liu, W. Research on interactive integrated planning of data center and distribution network driven by carbon emission reduction. *Trans. China Electrotech. Soc.* **2023**, *38*, 6433–6450. [CrossRef]
2. Stojanović, B.; Rajić, T.; Šošić, D. Distribution network reconfiguration and reactive power compensation using a hybrid Simulated Annealing–Minimum spanning tree algorithm. *Int. J. Electr. Power Energy Syst.* **2023**, *147*, 108829. [CrossRef]
3. Yan, X.; Zhang, Q. Research on Combination of Distributed Generation Placement and Dynamic Distribution Network Reconfiguration Based on MIBWOA. *Sustainability* **2023**, *15*, 9580. [CrossRef]
4. Mi, Y.; Chen, Y.; Yuan, M.; Li, Z.; Tao, B.; Han, Y. Multi-timescale optimal dispatching strategy for coordinated source-grid-load-storage interaction in active distribution networks based on second-order cone planning. *Energies* **2023**, *16*, 1356. [CrossRef]
5. Ma, C.; Wang, D.; Shan, X.; Wu, H.; Xu, X.; Wang, Y.; Guo, Y.; Zhang, J. A fault recovery strategy of distribution network based on mixed-integer second-order cone programming. In Proceedings of the 2020 5th Asia Conference on Power and Electrical Engineering (ACPEE), Chengdu, China, 4–7 June 2020; IEEE: Piscataway, NJ, USA, 2020; pp. 1584–1589. [CrossRef]
6. Kazemi-Robati, E.; Sepasian, M.S. Fast heuristic methods for harmonic minimization using distribution system reconfiguration. *Electr. Power Syst. Res.* **2020**, *181*, 106185. [CrossRef]
7. Essallah, S.; Khedher, A. Optimization of distribution system operation by network reconfiguration and DG integration using MPSO algorithm. *Renew. Energy Focus* **2020**, *34*, 37–46. [CrossRef]
8. Wang, B.; Zhu, H.; Xu, H.; Bao, Y.; Di, H. Distribution network reconfiguration based on NoisyNet deep Q-learning network. *IEEE Access* **2021**, *9*, 90358–90365. [CrossRef]
9. Li, Y.; Hao, G.; Liu, Y.; Yu, Y.; Ni, Z.; Zhao, Y. Many-objective distribution network reconfiguration via deep reinforcement learning assisted optimization algorithm. *IEEE Trans. Power Deliv.* **2021**, *37*, 2230–2244. [CrossRef]
10. Veličković, P.; Cucurull, G.; Casanova, A.; Romero, A.; Lio, P.; Bengio, Y. Graph attention networks. *arXiv* **2017**, arXiv:1710.10903. [CrossRef]

11. Liu, Z.; Liu, Y.; Xu, H.; Liao, S.; Zhu, K.; Jiang, X. Dynamic economic dispatch of power system based on DDPG algorithm. *Energy Rep.* **2022**, *8*, 1122–1129. [CrossRef]
12. Tavana, M.; Soltanifar, M.; Santos-Arteaga, F.J. Analytical hierarchy process: Revolution and evolution. *Ann. Oper. Res.* **2023**, *326*, 879–907. [CrossRef]
13. Xu, C.; Dong, S.; Zhu, J.; Zhu, B.; Xu, L. A radial constraint description method for distribution networks based on non connected power supply loop conditions. *Autom. Electr. Power Syst.* **2019**, *43*, 82–89. [CrossRef]
14. Trivić, B.; Savić, A. Optimal Allocation and Sizing of BESS in a Distribution Network with High PV Production Using NSGA-II and LP Optimization Methods. *Energies* **2025**, *18*, 1076. [CrossRef]
15. Bahrami, S.; Chen, Y.C.; Wong, V.W.S. *Dynamic Distribution Network Reconfiguration with Generation and Load Uncertainty*; IEEE: Piscataway Township, NJ, USA, 2024. [CrossRef]
16. Yang, X.; Wang, L. Probabilistic Power Flow Based Distribution Network Optimization with Distributed Generation. *Acta Energ. Solaris Sin.* **2021**, *42*, 71–76. [CrossRef]
17. Maran, D.; Olivieri, P.; Stradi, F.E.; Urso, G.; Gatti, N.; Restelli, M. Online markov decision processes configuration with continuous decision space. *Proc. AAAI Conf. Artif. Intell.* **2024**, *38*, 14315–14322. [CrossRef]
18. Cong, P.; Tang, W.; Lou, C.; Zhang, B.; Zhang, L. Two-stage coordinated optimal control of flexible soft open points and tie switches in active distribution networks with high penetration of renewable energy. *Trans. China Electrotech. Soc.* **2019**, *34*, 1263–1272. [CrossRef]
19. Shin, M.J.; Choi, D.H.; Kim, J. Cooperative management for PV/ESS-enabled electric vehicle charging stations: A multiagent deep reinforcement learning approach. *IEEE Trans. Ind. Inform.* **2019**, *16*, 3493–3503. [CrossRef]
20. Xihe. Energy Meteorological Big Data Platform [DB]. Available online: https://xihe-energy.com (accessed on 10 February 2025).

energies

MDPI

Article

Assessment of Grid-Tied Renewable Energy Systems' Voltage Support Capability Under Various Reactive Power Compensation Devices

Jie Cao [1], Mingshun Liu [1], Qinfeng Ma [1], Junqiu Fan [1], Dongkuo Song [2,*], Xia Zhou [3], Jianfeng Dai [3] and Hao Wu [3]

1 Guizhou Power Grid Co., Ltd., Guiyang 550002, China; caojie7293@163.com (J.C.); liumingshunedu7@163.com (M.L.); maqinfengous@163.com (Q.M.); fjqfjq73267@163.com (J.F.)
2 Guodian Nari Nanjing Control System Co., Ltd., Nanjing 210008, China
3 School of Automation, Nanjing University of Posts and Telecommunications, Nanjing 210023, China; zhouxia@njupt.edu.cn (X.Z.); daijianfeng@njupt.edu.cn (J.D.); harloworld@163.com (H.W.)
* Correspondence: songdongkuo7@163.com

Abstract

The weak grid strength in regions with large-scale renewable energy integration has emerged as a universal challenge, limiting the further expansion of renewable energy development. Currently, the short-circuit ratio (SCR) is widely used to quantify the relative strength between AC systems and renewable energy. To address this issue, this study first analyzes and compares how different reactive power compensation methods enhance the SCR. It then proposes calculation frameworks for both the SCR and critical short-circuit ratio (CSCR) in renewable energy grid-connected systems integrated with reactive power compensation. Furthermore, based on these formulations, a quantitative evaluation methodology for voltage support strength is developed to systematically assess the improvement effects of various compensation approaches on grid strength. Finally, case studies verify that reactive power compensation provided by synchronous condensers effectively strengthens grid strength and facilitates the safe expansion of the renewable energy integration scale.

Keywords: weak grid strength; renewable energy integration; voltage support strength; synchronous condensers

1. Introduction

The power system is progressively evolving into a "double high" system, characterized by a significant increase in renewable energy sources and a high percentage of electrically powered devices, as China's energy consumption shifts toward greener and low-carbon alternatives [1]. As the installed capacity of renewable energy continues to expand and the share of traditional thermal power decreases, the composition and operational modes of the power system have undergone substantial changes [2]. This transformation has resulted in a narrowing of the short-circuit current control margin for equipment operating at various voltage levels [3].

Only a few sizable power plants comprise the transmission end, while the majority of emerging energy sources, such as solar and wind, are located in remote areas [4]. These are AC systems at the transmission end with relatively weak strength due to their limited network structure, insufficient AC–DC support, and inadequate short-circuit capacity [5].

When the receiving system experiences a failure, it can lead to power surges, temporary overvoltages, and commutation errors in the DC network at the sending end [6]. Voltage instability, which has an immediate impact on both active and reactive power, along with other issues, may arise if a short-circuit fault occurs at the transmitting end of the network [7]. The strength of the network can be enhanced, and electrical stability ensured, by installing appropriate reactive power compensation devices at the converter station [8]. In the centralized grid connection generated by renewable energy, the SCR values fluctuate under different operating conditions of the system. The CSCR is the SCR at the critical stability of the system, indicating the lower limit value of the SCR during stable operation. In other words, the SCR value at the critical stability state is equivalent to the CSCR value. The greater the difference between the SCR and CSCR, the more stable the system becomes. By comparing these values, it is possible to assess the stability and safety of the developed renewable energy input system under the rated operating conditions [9]. This evaluation facilitates the optimization of the capacity of reactive power compensation devices [10].

Reference [11] evaluated the strength of power grids with high levels of inverter resource penetration by analyzing the short-circuit ratio (SCR). Reference [12] developed a small-signal model for grid-connected inverters, configuring them to operate at a leading power factor and assessing the system's operational status through variations in the short-circuit ratio. Reference [13] examined the stability of virtual synchronous machine grid-connected converters by characterizing the strength of AC systems based on the short-circuit ratio. Reference [14] utilized power-voltage curves to assess the constraints of single inverter terminal voltage while determining the minimum short-circuit ratio for an infinite bus system. Most current research tends to overlook the impact of reactive power-compensating devices, focusing instead on the evaluation of the SCR ratio in multi-infeed transmission networks.

Given the aforementioned issues, this study proposes a quantitative assessment method to calculate the short-circuit ratio (SCR) control margin of renewable energy power plants based on the computation of the SCR and CSCR of multiple renewable energy stations. This method, grounded in the direct current (DC) SCR, serves as the baseline for the constant capacity index. It selects the configuration with the lowest cost that meets the voltage strength requirements of the grid as the optimal setup for reactive power compensation devices [15], thereby guiding the operational management and planning of the renewable energy grids. First, the specific contributions of three reactive power compensation devices—the synchronous condenser, static var generator (SVG), and static var compensator (SVC)—to the system's voltage strength are thoroughly investigated. Using the SCR calculation model, the distinct methods by which each device enhances voltage support capacity under the same node configuration are compared in detail. Next, the short-circuit capacity of renewable energy grid-connected buses, the maximum reactive power compensation capacity, and the SCR of multiple stations with reactive power compensation are accurately calculated step by step. Subsequently, the underlying mechanisms by which these devices influence the SCR are analyzed in depth. The network's maximum transmission power is used to concurrently calculate the station's CSCR with precise reactive power adjustments. Then, the SCR control margin, derived from the difference between the SCR and CSCR, is computed to objectively evaluate the voltage stability of the new energy feed-in system. Finally, the enhanced contributions of the three devices to voltage stability are examined by quantifying the SCR control margin, and case studies are employed to comprehensively assess the actual performance of different devices. The results indicate that incorporating a synchronous condenser significantly improves the system's overall voltage support capability in practical scenarios.

2. Voltage Support Strength Enhancement Mechanisms in Renewable Grid-Connected Systems Under Diverse Compensation Devices

Currently, a popular method for reactive power compensation aimed at enhancing the weak grid system of energy clusters and improving the voltage support capability of newly developed energy network-connected systems involves the installation of static var compensators (SVCs), static synchronous generators (SVGs), and synchronous condensers [16]. Each reactive power adjustment strategy operates through a distinct mechanism, thereby strengthening the system's voltage support capability [17]. The next section will examine the impact of these three devices on the voltage stability of the network [18]. The short-circuit ratio (SCR) calculation model will be employed to compare the methods for enhancing the voltage support strength of each device within the same node configuration [19].

2.1. Static Var Compensator Enhancement Mechanism

Static var compensators (SVCs) are widely used compensation devices, and large-capacity power electronic devices have been extensively employed for reactive power compensation in power systems in recent years [20]. An SVC thyristor-controlled reactor (*TCR*) and a thyristor-controlled switching capacitor (*TSC*) often make up an SVC. The thyristor's conduction angle α can be adjusted to modify the equivalent impedance of a *TCR* and *TSC*. By injecting or absorbing reactive current into or out of the system, the SVC helps maintain system voltage and suppress oscillations, allowing it to quickly adapt to changes in the system during failures [21].

The reactive power output will adjust under variations in the bus voltage at the installation location, according to the SVC's operating principle. The equivalent impedance of *TCR* can be smoothly regulated by varying the thyristor's conduction angle α. The conduction angle and the equivalent susceptance value of TCR are related in the following way:

$$B_{eq} = \frac{\alpha - \sin\alpha}{\pi X_L} \tag{1}$$

In the formula, B_{eq} is the equivalent electronegativity, α is the conduction angle, and X_L is the sensor reactance. The inductive reactive power value injected into the grid by *TCR* is

$$Q_{TCR} = -B_{eq}U^2 \tag{2}$$

In the formula, U is the bus voltage at the SVC installation location.

*TSC*s are equivalent to capacitors that can be quickly switched, and the inductive reactive power injected into the system by each set of capacitors is

$$Q_{TSC} = \omega C U^2 \tag{3}$$

In the formula, ωC is the impedance value of the capacitor.

The inductive reactive power injected into the system by the SVC is

$$Q_{SVC} = U^2(\omega C - \frac{\alpha - \sin\alpha}{\pi X_L}) \tag{4}$$

In the formula, C is the filter capacitor, X_L is the reactance of the filter, ω is the voltage angular frequency, and α is the conduction angle of the thyristor.

Analyzing the impact of SVC input on system voltage intensity from the perspective of the short-circuit ratio index, after configuring the SVC at node x of the new energy field station, it is considered a shunt-connected capacitor and included in the admittance matrix.

When the SVC is connected to node x, it is equivalent to a parallel variable capacitor BSVC. Therefore, the short-circuit capacity of node x is

$$S_{ac,x} = U_N E_{eq,x} \mid Y'_{xx} \mid \quad (5)$$

In the formula, U_N is the nominal voltage at the grid connection point; $E_{eq,x}$ is the equivalent potential of the grid point x; Z_{ii} is the self-impedance of the grid-connected busbar of the new energy station x; and Y_{xx} is the self-admittance of the grid-connected busbar of the new energy station x.

Among them, the self-admittance Y_{xx} of node x after connecting the SVC is the following:

$$Y'_{xx} = G_{xx} + j(B_{xx} + B_{SVC}) \quad (6)$$

In the formula, B_{SVC} is the SVC susceptance; G_{xx} is the self-conductance of the grid-connected bus of the new energy station x; and B_{xx} is the self-susceptance of the grid-connected bus of the new energy station x.

The above formula can be converted into

$$S'_{ac,x} = U_N E_{eq,x} \sqrt{G_{xx}^2 + (B_{xx} + B_{SVC})^2} \quad (7)$$

$$MRSCR_x = \frac{S'_{ac,x}}{P_{N,x}} \quad (8)$$

In the formula, $P_{N,x}$ is the rated capacity of the new energy grid connection point x.

Following the calculation of the short-circuit ratio, it can be inferred that the value MRSCR will increase in conjunction with the rise in the Y_{xx} modulus once node x is configured with the SVC. This adjustment elevates the system's SCR value, thereby enhancing the voltage support level and strengthening the system's resilience to voltage fluctuations. The short-circuit ratio (SCR) characterizes the voltage support strength of the renewable power grid. A high SCR indicates both the robustness of the network and the system's ability to maintain voltage stability. By activating the capacitor bank, the system's SCR can be increased, thus improving the overall voltage stability of the system.

2.2. Static Var Generator Enhancement Mechanism

One of the key components of Flexible AC Transmission Systems (FACTS) is the static var generator (SVG). The SVG offers several advantages, including its compact size, rapid response time, and wide operating range [22]. It can detect fluctuations in the reactive power of renewable energy systems with high sensitivity and promptly compensate for these changes [23]. In addition to reactive power compensation, the SVG effectively suppresses harmonic currents, reduces network losses, and enhances power quality. It also plays a crucial role in controlling system voltage fluctuations and improving overall system robustness [24]. Due to these benefits, SVGs have been extensively studied and widely implemented [25].

The power demand of the load and the system's capacity to supply power to that load define the transient voltage stability of the developed renewable power system. The Thévenin equivalent system, as illustrated in Figure 1, offers a straightforward way to understand this concept.

In Figure 1, E and Z_e represent the equivalent potential and impedance on the system side, respectively; U_l is the voltage of the load node; Z_l and I_l represent load impedance and load current, respectively; I_{SVG} is the compensating current for the SVG; and I_{eq} is the equivalent load current after SVG connection.

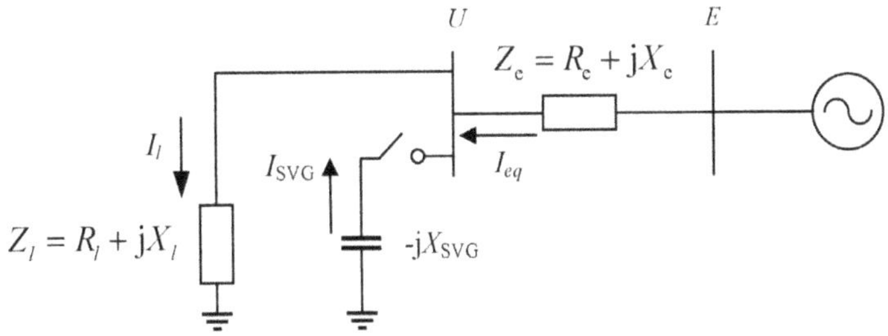

Figure 1. The Thévenin equivalent system for the SVG access system.

Based on the derivation of the Thévenin equivalent circuit shown in the figure above, when the SVG is not connected to the system, the expressions for active and reactive power in the parallel line can be calculated, and the voltage at the load node can be determined as follows:

$$U_l = \left[\frac{E^2}{2} - P_l R_e - Q_l X_e \pm \sqrt{(P_l R_e + Q_l X_e - \frac{E^2}{2})^2 - Z_e^2(P_l^2 + Q_l^2)}\right]^{\frac{1}{2}} \tag{9}$$

In a steady state, the modulus of the load impedance, $|Z_l|$, is greater than the modulus of the system's equivalent impedance, $|Z_e|$. Assuming that the load's power factor remains constant throughout the transient process, the load increases its power by reducing its equivalent impedance, causing its impedance modulus to decrease from a larger value to a smaller one. The system ultimately operates at the intersection of the upper and lower halves of the PV curve, resulting in a unique voltage solution when $|Z_l| = |Z_e|$. At this load node, the voltage is in a critically stable state. However, at this point, the voltage becomes unstable if $|Z_l|$ continues to decrease.

After the system is integrated into the static var generator (SVG), based on the principle of circuit equivalence, the SVG, supported by capacitors on the DC side, can be considered an adjustable capacitor C_{SVG}. When observing from the load node, the SVG and load impedance can be represented as an equivalent impedance Z_{eq}. The relationship between the current I_{eq} flowing through Z_{eq}, the SVG compensation current I_{SVG}, and the load current I_l is as follows:

$$\dot{I}_{eq} = \dot{I}_l - \dot{I}_{SVG} \tag{10}$$

The equivalent impedance Z_{eq} of the load node after incorporating the SVG can be obtained from the current relationship, and the expression is

$$Z_{eq} = \frac{X_l X_{SVG} - \mathbf{j} R_l X_{SVG}}{R_l + \mathbf{j}(X_l - X_{SVG})} \tag{11}$$

In the equation, X_{SVG} is the equivalent reactance of the SVG.

The partial derivative of X_{SVG} for $|Z_{eq}|^2$ is obtained as follows:

$$\frac{\partial |Z_{eq}|^2}{\partial X_{SVG}} = \frac{2X_{SVG}(R_l^2 + X_l^2)[R_l^2 + X_l^2 - X_l X_{SVG}]}{[R_l^2 + (X_l - X_{SVG})^2]^2} \tag{12}$$

Because there is $R_l \ll X_l$ in the high-voltage transmission system, it can be assumed that $R_l \approx 0$, and the above formula can be simplified as

$$\frac{\partial |Z_{eq}|^2}{\partial X_{SVG}} = \frac{2X_{SVG}(X_l^2)[X_l^2 - X_l X_{SVG}]}{(X_l - X_{SVG})^4} \tag{13}$$

During the system's temporary operation, as the equivalent capacitance C_{SVG} of the static var generator (SVG) increases, the corresponding reactance X_{SVG} decreases. Furthermore, since $I_l > I_{SVG}$, it can be inferred from the monotonicity of the function that the magnitude $|Z_{eq}|$ of the compensated equivalent load impedance increases as X_{SVG} decreases, thereby enhancing the system's dynamic voltage stability.

After the SVG is configured at the node x of the renewable energy station, it is treated as a separate current source and included in the short-circuit ratio calculation. This analysis evaluates the impact of the SVG input on system voltage levels from the perspective of short-circuit ratio indicators. The short-circuit capability at the node x is

$$S_{ac,x} = U_N E_{eq,x} \mid Y'_{xx} \mid \tag{14}$$

In the formula, U_N represents the nominal voltage at the grid connection point; $E_{eq,x}$ denotes the equivalent potential of grid point x; Z_{ii} indicates the self-impedance of the grid-connected busbar at the new energy station x; and Y_{xx} signifies the self-admittance of the grid-connected busbar at the new energy station x.

Among them, the node x is connected to the SVG, and the equivalent admittance $Y_{xx}$is

$$Y'_{xx} = G_{xx} + j(B_{xx} + \frac{Q_{SVG}}{U_N E_{eq,x}}) \tag{15}$$

In the formula, G_{xx} represents the self-conductivity of the grid-connected busbar of the new energy station x, B_{xx} denotes the self-electrification of the grid-connected busbar of the new energy station x, and Q_{SVG} indicates the injected reactive power.

The above formula can be changed into the following by replacing it in the short-circuit capacity calculation formula:

$$S'_{ac,x} = U_N E_{eq,x} \sqrt{G_{xx}^2 + \left(B_{xx} + \frac{Q_{SVG}}{U_N E_{eq,x}}\right)^2} \tag{16}$$

For the sake of convenience in calculations, and assuming that different DC systems do not interfere with one another, the formula for calculating the SCR ratio of the x-th node in the new energy system after the SVG layout can be derived as follows:

$$MRSCR_x = \frac{S'_{ac,x}}{P_{N,x}} \tag{17}$$

The rated capacity of the grid connection point x for new energy is denoted by $P_{N,x}$ in the formula.

After analyzing the calculation of the SCR, it can be concluded that after configuring the SVG at the node x, the value of the MRSCR increases as the modulus $|Y_{xx}|$ increases. This change results in a higher SCR ratio, thereby enhancing the system's resistance to voltage fluctuations and improving the voltage support level. The SCR ratio is a key indicator of voltage support strength; a higher SCR ratio signifies a stronger network and greater capacity for maintaining voltage stability. By implementing the SVG, the system's SCR ratio can be elevated, which contributes to improved voltage stability.

2.3. Synchronous Condenser Enhancement Mechanism

Synchronous condensers were first utilized as reactive power compensation devices for the electrical grid around the turn of the 20th century. However, more advanced reactive power compensation technologies have progressively replaced them. Many countries and regions have reintroduced condensers due to their unique advantages in enhancing frequency and inertia support [26]. The conversion of decommissioned thermal power units into condensers has also gained significance as a means to ensure grid stability [27]. Synchronous condensers are rotating devices capable of compensating for dynamic reactive power [28]. Their effective reactive power output characteristics can bolster the system's inertia and short-circuit capacity as well as maintain the system voltage and improve the overall system stability [29]. Consequently, synchronous condensers with specific capacities are often installed near renewable energy sources in engineering applications [30].

As shown in Figure 2, after the development of the renewable energy station, node x configured with a condenser can be considered to have a grounding branch with impedance Z_{xSC} added to it.

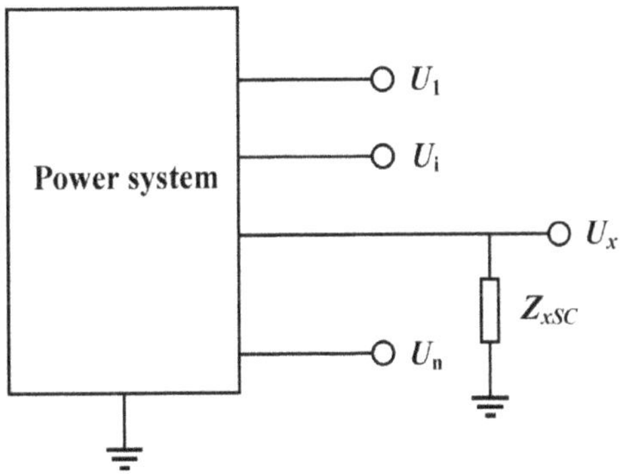

Figure 2. Influence of synchronous condenser access on node impedance matrix.

According to the principle of the additional branch method, it can be deduced that the impedance value between node i and node j will change accordingly:

$$Z'_{ij} = Z_{ij} - \frac{Z_{xi}Z_{xj}}{Z_{xx} + Z_{xSC}} \tag{18}$$

The AC and DC axis voltages are represented by u_q and u_d, respectively, while the currents are denoted by I_q and I_d. Z_{xi} refers to the element in the x-th row and i-th column of the equivalent impedance matrix of the AC power grid at the energy grid-connected busbar. Similarly, Z_{xj} represents the element in the x-th row and j-th column of the same impedance matrix. Z_{xx} indicates the self-impedance of the grid-connected busbar at energy station x. Lastly, Z_{ij} denotes the element in the i-th row and j-th column of the equivalent impedance matrix of the AC power grid at the grid-connected busbar.

After node x is configured with the synchronous condenser, the equivalent impedance matrix of the system will show a decrease in all its elements. The node most affected is node x, which is integrated with the synchronous condenser; specifically, Z_{xx} experiences the most significant reduction.

By thoroughly integrating the short circuit ratio calculation method and assuming that the voltage phase angle between each station is similar, as well as that the voltage at each node is close to the rated value, the SCR calculation formula for the i-th branch after the capacitor arrangement can be derived as follows:

$$MRSCR = \frac{U_{N,i}^2}{\sum_{j=1}^{n} |Z_{ij}| P_j} \tag{19}$$

The rated voltage of energy station i's grid-connected busbar is denoted as $U_{N,i}$, and the active power produced by the station is represented by P_i. The term Z_{ij} refers to the i-th row and j-th column element of the equivalent impedance matrix of the AC power grid at the energy grid-linked busbar, while P_j denotes the active power produced by energy station j.

After node x is integrated into the synchronous condenser, it can be determined through the SCR ratio calculation that the MRSCR incorporated into the condenser node is significantly influenced by the modulus value $|Z_{ij}|$. Specifically, the value of MRSCR increases as the $|Z_{ij}|$ modulus decreases. This enhancement improves the SCR ratio of the network, thereby bolstering its resistance to voltage fluctuations and elevating the voltage support level. A high SCR ratio indicates the network's strength and its capacity to maintain voltage stability. Conversely, a low SCR ratio suggests instability within the system and indicates a weak network. Consequently, the condenser is activated in the event of a system failure, which can raise the system's SCR ratio and improve overall stability.

It is evident from the SCR ratio calculation process outlined above that the availability of various reactive power compensation techniques will influence the SCR ratio calculation at each grid-connected point. This, in turn, will affect the voltage-sustaining capacity of each grid-dependent point. The SVC is considered to be the admittance matrix of the bus in parallel with the bus, which explains why the compensation effect of the synchronous condenser is superior to that of the SVC. Among these, the inputs from the SVC and synchronous condenser primarily impact the short-circuit capacity at every point in the system by modifying the network's admittance matrix. However, the integration of the synchronous condenser directly affects the self-admittance at the access node's location. The input from the SVG is primarily utilized as a separate power source to contribute to the calculation of the short-circuit ratio, thereby enhancing the reactive power at the system nodes.

This section establishes the computational and physical models of various reactive power adjustment devices connected to the developed renewable energy station. It also compares the effects of reactive power compensation and the support capacity of different types of reactive power compensation devices on network voltage. Theoretically, the condenser is shown to be the most effective option for adaptive power compensation. To enhance the SCR index and improve transmission efficiency, the new energy station is equipped with a distributed condenser.

3. Quantitative Assessment of Voltage Support Strength in Grid-Connected Renewable Energy Stations

Conventional power supply systems lack damping support and inertia, connecting to the network as grid-following current sources. Consequently, they are unable to actively respond to variations in grid voltage and frequency [31]. As a result, conventional AC systems must maintain their connections through voltage and frequency regulation with the strength of the AC system—relative to its electrical equipment—primarily determining the overall system's stability [32]. Currently, significant amounts of developed renewable energy are integrated into the network via power electronic inverters [33]. Numerous researchers have examined the capacity of energy grid-dependent systems and proposed the concepts of voltage support strength and frequency support strength to characterize these systems' robustness [34]. Among these, frequency support strength, typically assessed using indicators such as the frequency change rate and the inertia constant, reflects the dampening effect of frequency fluctuations caused by system disturbances [35]. Voltage support strength, on the other hand, is generally evaluated through indicators like the SCR and the impedance ratio, illustrating the system's ability to maintain stable voltages following disruptions under specific initial operational conditions.

The CSCR ratio algorithm ensures a consistent integration of calculation methods and criteria for determining system strength, serving as a crucial reference for evaluating developed renewable energy input systems. The concept of system voltage support strength, while abstract, is grounded in the classification of system strengths and weaknesses [36]. By utilizing the critical short-circuit ratio as a reference point and establishing a coordinate system based on the short-circuit ratio index, this approach effectively assesses the stability of the developed renewable energy input system [37].

3.1. Short-Circuit Ratio Index for Voltage Support Strength of Renewable Energy Power Stations Under Reactive Power Compensation

Increasing the amount of reactive power support at the receiving end is a standard method for enhancing the receiving system's ability to maintain voltage stability [38]. Parallel static capacitors are commonly employed as reactive power sources due to their capacity to optimize the network's energy distribution [39]. When the system operates steadily, the internal reactive power consumption is balanced with the reactive power supplied by the reactive power source [40]. However, the reactive electrical energy produced by parallel capacitors diminishes in conjunction with the voltage drop caused by a fault in the receiving electrical network [41]. In certain emergency scenarios, parallel capacitors can adversely affect system voltage stability. In these instances, dynamic power compensation mechanisms are necessary to enhance voltage stability, as parallel capacitors are not suitable for the network's transient operating conditions.

The equivalent model of the renewable energy grid connection system, considering reactive power compensation devices, is shown in Figure 3.

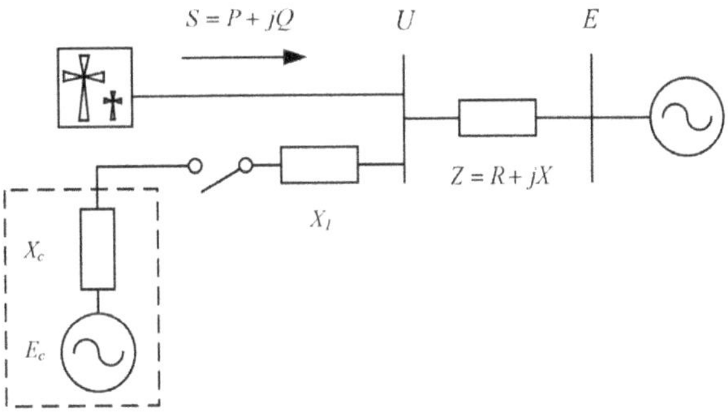

Figure 3. Equivalent circuit diagram of a reactive power compensation device connected to a renewable energy grid connection system.

The short-circuit capacity of the grid-connected bus in the power station with a reactive power capacity, taking into account the effect of the paralleled reactive power adjustment capacity $Q_{s,i}$ on the DC converter bus, is

$$S_{vc,i} = \frac{U_N E_{eq,i}}{Z_{ii}} - Q_{s,i} \tag{20}$$

where $E_{eq,i}$ is the equivalent potential of node i, and U_N is the nominal voltage at the junction point. The developed renewable energy station i's networked bus has a reactive power capacity of $Q_{s,i}$ and a self-impedance of Z_{ii}.

The reactive electricity compensation device SCR ratio for developed renewable energy single stations is

$$SCR_i = \frac{S_{vc,i}}{P_{N,i}} \tag{21}$$

where $P_{N,i}$ is the rated capacity of the developed renewable energy station in network i; and $S_{vc,i}$ is the short-circuited capacity of the grid-connected connector bar of the renewably constructed power plant with the SVC reactive power compensation device.

The DC landing locations in a multi-infeed DC system are comparable, and interference between multiple DC systems is possible. When evaluating the planning scheme of a multi-infeed DC system using only the SCR, the findings are often biased toward more ideal conditions. This paper utilizes the short-circuit capacity and reactive power compensation capacity of renewable energy grid-connected busbars to calculate the short-circuit ratio of various developed renewable energy systems in which the generators are equipped with reactive power compensation devices.

Based on the Thévenin equivalent impedance, the node voltage is

$$\begin{bmatrix} \dot{U}_{S,1} \\ \dot{U}_{S,2} \\ \cdots \\ \dot{U}_{S,i} \\ \cdots \\ \dot{U}_{S,n} \end{bmatrix} = \begin{bmatrix} Z_{11} & Z_{12} & \cdots & Z_{1i} & \cdots & Z_{1n} \\ Z_{21} & Z_{22} & \cdots & Z_{2i} & \cdots & Z_{2n} \\ \cdots & \cdots & \cdots & \cdots & \cdots & \cdots \\ Z_{i1} & Z_{i2} & \cdots & Z_{ii} & \cdots & Z_{in} \\ \cdots & \cdots & \cdots & \cdots & \cdots & \cdots \\ Z_{n1} & Z_{n2} & \cdots & Z_{ni} & \cdots & Z_{nn} \end{bmatrix} \begin{bmatrix} \dot{I}_{S,1} \\ \dot{I}_{S,2} \\ \cdots \\ \dot{I}_{S,i} \\ \cdots \\ \dot{I}_{S,n} \end{bmatrix} \tag{22}$$

$U_{S,i}$ in the formula represents the station's grid-tied transit voltage; at the renewably installed power grid-connected busbar, Z_{ij} is the i-th row and j-th column member of the AC power grid's equivalent impedance matrix; $I_{S,i}$ injects current into the renewably constructed power AC network; and Z_{ii} is the self-impedance of the electrically linked busbar.

The short-circuiting potential of the electrically linked bus in renewably constructed energy plants with reactive power adjustment equipment is evaluated in light of the effect of parallel dynamic power adjustment capabilities $Q_{s,i}$ on the DC converter bus:

$$S_{vc,i} = \left|\dot{U}_{N,i}\dot{I}_{s,i}^*\right| - Q_{s,i} = \left|\frac{\dot{U}_{N,i}\dot{U}_{S,i}^*}{Z_{ii}}\right| - Q_{s,i} \tag{23}$$

* represents the conjugate operation in the formula. $I_{S,i}$ is the short-circuiting current at the grid-connected bus bar of the renewably constructed energy station; $U_{N,i}$ is the rated voltage of the renewably constructed power station's networked bus; $Q_{s,i}$ is the reactive power adjustment capability; $U_{S,i}$ is the grid bus voltage connected to the renewably constructed power plant; and Z_{ii} is a representation of the self-impedance of the electrically connected bus of the station i.

Reactive power offset for the device short-circuiting ratio of a renewably constructed energy multi-station:

$$MRSCR_i = \frac{S_{vc,i}}{\left|S_{S,i} + \sum_{j=1,j\neq i}^{n} \varepsilon_{ij} S_{S,j}\right|} \tag{24}$$

$S_{S,i}$ and $S_{S,j}$ are the real seeming strength of the renewably generated energy injected into the i and j grid-tied bus terminals of renewably constructed energy; and $S_{vc,i}$ is the short-circuiting capacity of the electrically linked bus bar of the renewably constructed energy station with reactive power compensation devices.

In the above formula, ε_{ij} is an indicator for measuring the voltage coupling relationship at the receiving end between multi-feed systems, which is used to characterize the interaction intensity between converter stations in multi-feed transmission systems:

$$\varepsilon_{ij} = \frac{Z_{ij} U_i^*}{Z_{ii} U_j^*} \tag{25}$$

The following formula can be derived from the renewably constructed power production station's electrically linked bus's short-circuiting capacity:

$$MRSCR_i = \frac{\left|\dot{U}_{N,i}\dot{U}_{S,i}^*/Z_{ii}\right| - Q_{s,i}}{\left|\dot{U}_{S,i}^*\dot{I}_{S,i} + \sum_{j=1,j\neq i}^{n} \frac{Z_{ij}}{Z_{ii}} \dot{U}_{S,i}^* \dot{I}_{S,j}\right|} \tag{26}$$

$U_{N,i}$ is the rated voltage of the renewably constructed energy station i's electrically linked bus bar; $U_{S,i}$ is the electricity station i's grid-connected bus voltage; Z_{ii} is the electricity station i's grid-connected bus's self-impedance; $I_{S,i}$ and $I_{S,j}$ inject current into the AC system for units i and j; $Q_{s,i}$ is the reactive electricity adjustment capability; and Z_{ij} is the equivalent impedance between the grid-connected bus lines of stations i and j.

3.2. Calculation Method of CSCR for Renewable Power Stations with Reactive Compensators

Static steady voltage and the SCR ratio are interconnected through the CSCR ratio. The criterion for evaluating the stability of static voltage is the maximum transmission power. As the transmission power increases, the operating point shifts from the upper half of the P-V curve to the lower half, reaching the highest possible transmission energy at the curve's inflection point [42]. At this moment, the system is in a critically stable condition, and the CSCR ratio corresponds to the SCR ratio. Figure 4 illustrates a comparable circuit for the reactive electrical compensating device, which is designed to enhance the voltage performance of the developed renewable energy grid-connected network [43].

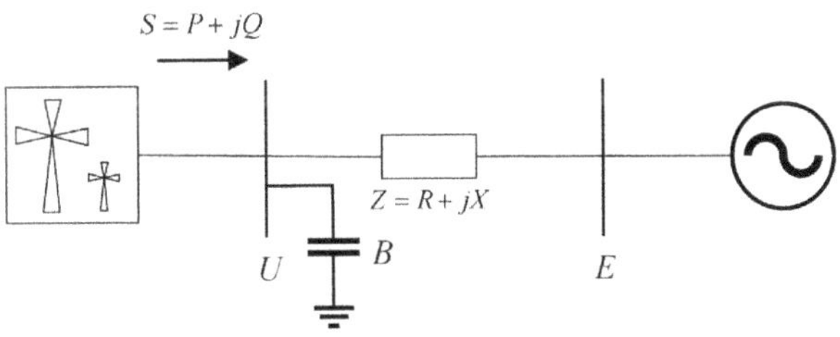

Figure 4. Equivalent model of renewable energy station with reactive power compensation device.

This section focuses on the ultimate reactive power output capability of various reactive power compensation devices, rather than their dynamic adjustment characteristics during real-time operation, such as response speed, adjustment accuracy, and output strategies under different operating conditions. The primary reason for this approach is that when evaluating the robustness of the system—specifically, its voltage support capability in response to extreme faults or significant disturbances—the maximum reactive power that a compensation device can provide directly reflects its potential to support the system when the voltage is on the verge of instability. At the same time, to achieve a unified analysis and comparison of different types of reactive power compensation devices, it is essential to consider the following: whether a static var compensator (SVC) relies on the phase control of thyristors, a static var generator (SVG) utilizes fully controlled power electronic devices for commutation and reactive power regulation, or a synchronous condenser depends on the electromagnetic characteristics of rotating motors to provide reactive power. The model must eliminate the differences arising from specific technical details and establish a standardized quantitative benchmark. Consequently, the limiting effect of all devices is characterized by their maximum reactive power capacity, denoted as $Q_{s,i}$, which is further transformed into equivalent susceptance B and integrated into the system model.

$$Q_{s,i} = U^2 B \tag{27}$$

The reactive electricity compensating point voltage is denoted by U, the reactive electricity compensating capacities by $Q_{s,i}$, and the capacitance to ground by B.

A quadratic equation of the connecting point's voltage can be derived from the system's highest transfer capacity:

$$U^4\beta - U^2[E^2 + 2(PR + QX - \gamma)] + S^2Z^2 = 0 \tag{28}$$

$$U = E\sqrt{\frac{1 + 2\lambda_s \pm \sqrt{\Delta_s}}{2\left[(1 - BX)^2 + B^2R^2\right]}} \tag{29}$$

$$\Delta s = 1 + 4\lambda_s - 4\mu_s^2 \tag{30}$$

Among them, $\lambda_s = \frac{PR+QX-QBZ^2}{E^2}$, $\mu_s = \frac{PX-QR-PBZ^2}{E^2}$, $\gamma = QBZ^2$, and $\beta = 1 - 2BX + B^2Z^2$. Reactive power compensated calculation factors are represented by γ and β in the formula, while λ_s and μ_s are calculation factors.

When equation $\Delta_s = 0$ is satisfied, the network's maximal capacity for transmission is

$$P_{S\max} = \frac{\left[R(E^2 + 2QX - 2\gamma) + EZ\sqrt{(E^2 + 4QX - 4\gamma)\beta}\right]}{2(X - BZ^2)^2} \tag{31}$$

$$U_{\text{Bcri}} = \sqrt{\frac{E^2 + 2(PR + QX - \gamma)}{2\beta}} \tag{32}$$

R is the resistance of the grid-connected bus of the renewably constructed energy plant station, and E is its potential. Q represents the reactive power of the energy grid-tied nodes, X represents the reactance at the energy plant station's grid-connected bus, γ and β represent the reactive power adjustment calculation factors, and B represents the ground capacitance.

Consequently, the following is the derivation of the renewable energy station's CSCR ratio with reactive power compensation equipment.

$$CSCR_i = \frac{S_{ac}}{|\ P_{S_{\max}} + jQ\ |} \tag{33}$$

The formula defines $P_{S\max}$ as the maximum transmission power of the developed renewable electrically linked system and S_{ac} as the grid-tied short-circuiting capacity of the renewably constructed energy plant.

3.3. Quantitative Analysis of Voltage Support Strength in Grid-Tied Renewable Energy Stations

The stability of renewable energy feeding systems can be quantitatively assessed by calculating the SCR margin for renewable energy stations equipped with reactive power adjustment devices. This assessment is based on the difference between the SCR ratio and the CSCR ratio. The stability margin of the SCR is evaluated by the relative magnitude of the difference between the SCR and the CSCR. The stability margin for a single-station system is denoted as η_{SCR}, while the stability margin for a multi-station system is represented by η_{MRSCR}.

Taking into account a reactive voltage adjustment device, the renewably constructed power individual station's short-circuiting margin is

$$\eta_{SCR} = \frac{SCR_i - CSCR_i}{CSCR_i} \times 100\% = \frac{\frac{S_{vc,i}}{S_{ac,i}}|P_{Smax} + jQ_i| - P_{N,i}}{P_{N,i}} \times 100\% \tag{34}$$

Q_i is the equivalent of reactivity that the additional energy transmits to the AC system, and j is an imaginary number in the formula. The nominal capacity of the renewably generated electricity at the connecting point i is denoted by $P_{N,i}$.

At the reactive voltage compensation device's grid connection point, take into account the SCR ratio margin of several renewable electricity stations:

$$\begin{aligned} \eta_{\mathrm{MRSCR}} &= \frac{\mathrm{MRSCR}_i - \mathrm{CSCR}_i}{\mathrm{CSCR}_i} \times 100\% \\ &= \frac{\frac{S_{\mathrm{vc},i}}{S_{\mathrm{ac},i}}|P_{\mathrm{Smax}} + jQ_i| - \left|S_{\mathrm{S},i} + \sum_{\substack{j=1 \\ j\neq i}}^{n} \varepsilon_{ij} S_{\mathrm{S},j}\right|}{\left|S_{\mathrm{S},i} + \sum_{\substack{j=1 \\ j\neq i}}^{n} \varepsilon_{ij} S_{\mathrm{S},j}\right|} \times 100\% \end{aligned} \tag{35}$$

Q_i is the equivalent reactive power that the renewably constructed energy sends to the AC framework; $S_{\mathrm{vc},i}$ is the short-circuiting capacity of the electrically linked bus of the installed energy station with a reactive power mitigation device; j is an imaginary number; and $S_{\mathrm{ac},i}$ is the short-circuiting capacity of the electrically linked bus line of the installed power station i. The symbols $S_{S,i}$ and $S_{S,j}$, respectively, represent the actual apparent power of new energy injected into the electrically linked bus nodes i and j.

The stability of the system is determined by comparing the SCR and CSCR ratios. The network's safety margin before reaching critical stability is indicated by the difference between the SCR and CSCR ratios.

1. When $\mathrm{SCR}_i < \mathrm{CSCR}_i$, or $\eta_{\mathrm{SCR}} < 0$, the developed renewable power grid-connected system for the single-feed system is unstable. The stability margin η_{SCR} increases with system instability; a system is in a stable condition when $\mathrm{SCR}_i > \mathrm{CSCR}_i$, which means $\eta_{\mathrm{SCR}} > 0$. The stability margin increases with system security.
2. In the case of the new energy multi-feed system, when $\mathrm{MRSCR}_i < \mathrm{CSCR}_i$, meaning that $\eta_{\mathrm{MRSCR}} < 0$, the system is unstable; the more unstable the system, the larger the stability margin η_{MRSCR}; when $\mathrm{MRSCR}_i > \mathrm{CSCR}_i$, meaning that $\eta_{\mathrm{MRSCR}} > 0$, the system is stable; the more stable the system, the larger the stability margin η_{MRSCR}.

4. Example Analysis

4.1. Stability Assessment of the System After Reactive Power Compensation

Construct a PSD-BPA-based model, determine a system's maximal power generation using the continuous power flow method, and create simulation examples of the system operating at various reactive power adjustment capacities. To quantitatively assess the viability and efficacy of the voltage support strength of the developed renewable energy feeding system, confirm the SCR ratio margin of renewably constructed energy stations with reactive voltage compensation devices, which is determined by the difference between the SCR ratio index and the CSCR ratio.

In the example of a single feed-in system, the parameter R is set to 0, the equivalent impedance of the AC system is $Z = 0.2j\,/\text{pu}$, the short-circuit capacity is $S_{\text{ac}} = 5\,/\text{pu}$, the reactive power is $Q = -0.4\,/\text{pu}$, the $P - V$ corresponds to the active power of the system $P = 2.06\,/\text{pu}$, and the operating voltage $U = 0.66\,/\text{pu}$. Figure 5 displays the outcomes of calculating various SCR ratio indicators for the system's critical steady state under varying compensated reactive power levels.

Reactive power compensation increases the maximum transmission power P_{max} when calculating the SCR ratio index in a single feed-in network. The ratio of the SCR capacity to the maximum power transfer is the formula used to calculate the CSCR. The shift in the system's maximum transmission power is the primary cause of the decline in the CSCR. Simultaneously, the rapid decrease in the CSCR encourages the improvement of the SCR ratio control margin η_{SCR}, enhances the system's ability to operate steadily in a steady state, and broadens the acceptance of incoming energy.

The interaction among various renewable energy sources is illustrated through the example of a multi-infeed plant. The AC system adheres to this equivalency, as shown in Table 1, which presents the relevant parameters. After assessing the renewable energy capacity at grid connection points 2 and 3, the $P - V$ curve for grid connection point 1 at a reactive power level of $Q = 0$ indicates an active power P of 2.5 pu and an operating voltage U of 0.707 pu.

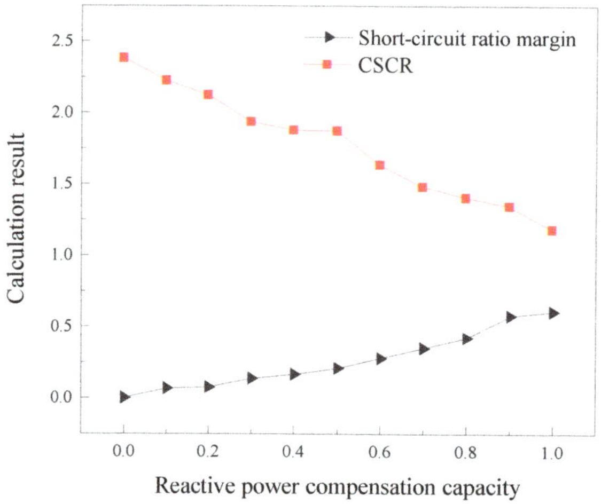

Figure 5. Short-circuit ratio related indicators of single feed-in system.

Table 1. Multiple input case parameters.

AC System Impedance	Branch 1	Branch 2	Branch 3
0.3*j*	0.12*j*	0.8*j*	0.8*j*

The changes in the parameters related to the SCR ratio of each grid connection point after compensating for different reactive power capacities are shown in Table 2, and the changes in the SCR ratio control margin are shown in Figure 6.

Due to the identical impedance of branches 2 and 3, as well as their equal grid-connected capacity, the multi-infeed equivalent model indicates that when the reactive power output, impedance parameters, and grid-connected capacity of the two infeed points are relatively symmetrical, their SCR ratio and CSCR ratio parameters will also be consistent. The computation results demonstrate that when compensation equipment is added, the CSCR decreases while the η_{MRSCR} increases. The new energy station access system remains stable when $\eta_{\mathrm{MRSCR}} > 0$. The greater the stability margin η_{MRSCR} of the system, the higher the transmitted power capacity it can accommodate. There are prerequisites for appropriately expanding the new energy input scale.

Table 2. SCR index of multi-feed-in system under different reactive power compensation capacities.

$Q_{s,i}$/pu	Node	MRSCR	CSCR	η_{MRSCR}
$Q_{s,i} = 0$	1	2.085	2.055	1.46%
	2	1.893	1.891	0.11%
	3	1.893	1.891	0.11%
$Q_{s,i} = 0.2$	1	2.008	1.852	8.42%
	2	1.817	1.803	0.78%
	3	1.817	1.803	0.78%
$Q_{s,i} = 0.4$	1	1.911	1.663	14.91%
	2	1.723	1.697	1.53%
	3	1.723	1.697	1.53%

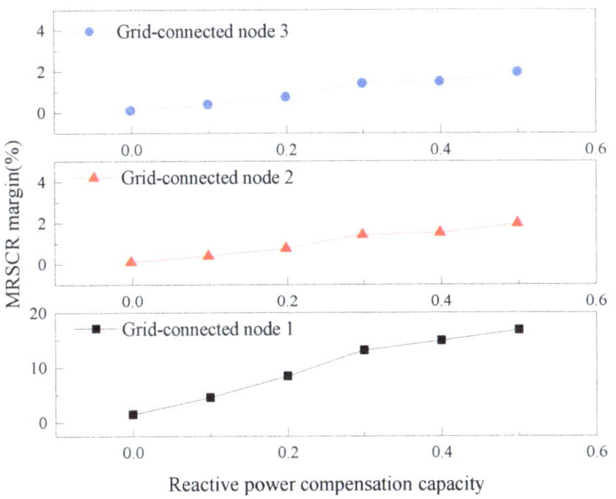

Figure 6. SCR ratio related indicators for multi feed-in systems.

4.2. Comparative Analysis of Renewably Constructed Energy System Stability Under Different Reactive Power Compensation Devices

Increasing the electrical intensity of renewable energy grid-connected points is the most effective method for enhancing the voltage performance of newly installed energy grid systems and increasing the electricity consumption capacity of weak current grid systems. This section examines the impact of various reactive power compensation systems on the power consumption generated by renewable energy stations. The engineering example system is illustrated in Figure 7. There are 81 renewable energy stations in the energy collection area, which are aggregated into four 500 kV substations and transmitted through a 1050 kV high-voltage station via UHV DC. The total installed capacity of new energy is approximately 12,000 MW, of which the installed capacity of wind power is about 8500 MW and the installed capacity of photovoltaics is 3500 MW. The grid structure of each DC access point in the receiving end of the AC/DC hybrid power grid is depicted in Figure 7.

To quantitatively evaluate the impact of these devices on the strength of the system's electrical network, an assessment of the improvement effects of various reactive electrical adjustment devices on the MRSCR was conducted using an experimental energy collection and transmission system as a model. It is important to note that the reactive electrical output of the SVC is 10% of its installed capacity when configured for compensation. When the SVG is configured, its reactive electrical output also equals 10% of the active electrical output of the supply. The reactive power output of the synchronous compensator is approximately 80% of its installed capacity when operational. Table 3 presents several sets of MRSCR index data for various reactive power compensation devices, while Figure 8 illustrates the effects of these devices on the improvement in the MRSCR.

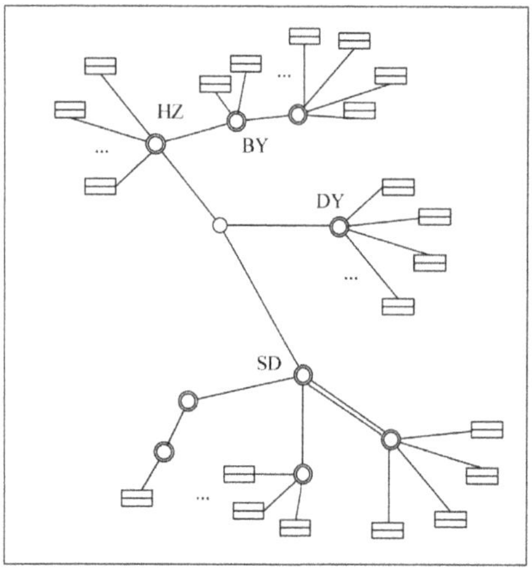

Figure 7. Actual grid structure of a certain confluence station in the power grid.

Table 3. Comparison of MRSCR indexes of each station under different reactive power compensation devices.

Renewable Energy Power Plant	Original System	SVC	SVG	Synchronous Compensator
SD	1.806	1.851	1.951	2.474
DY	1.880	1.948	2.128	3.005
HZ	1.981	2.041	2.137	3.126
BY	1.874	1.909	2.072	3.183

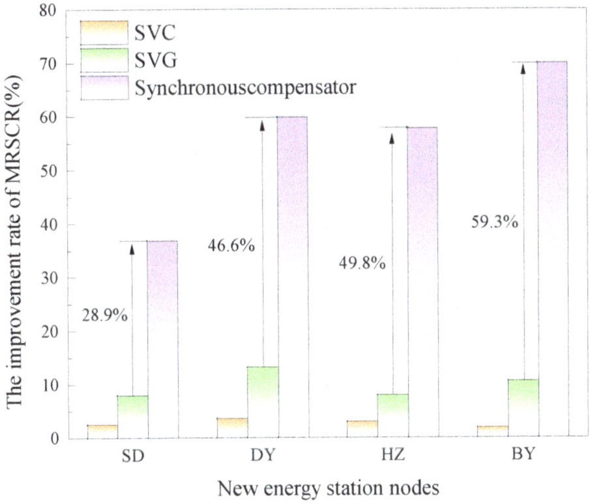

Figure 8. Comparison of the effects of different reactive power compensation devices on MRSCR improvement.

The analysis of the computational results indicates that the SCR ratio at the newly constructed grid-tied node can be improved by utilizing three reactive electrical compensation devices. Among these, the SVC exhibits the smallest increase in the MRSCR because it adjusts each node's short-circuit capacity. The increase rate varies from 1% to 10% with the majority of cases falling below 5%. In the MRSCR computation, the SVG is represented as a single current source. The increase in MRSCR ranges from 5% to 15% with SVG compensation slightly exceeding that of SVC. Access to the synchronous condenser reduces both the mutual and self-impedances of the node, thereby directly enhancing the MRSCR value. Consequently, the impact of the synchronous condenser on MRSCR enhancement is the most significant. The newly constructed renewable energy power plant, connected to the collection unit with the distributed synchronous condenser, has experienced an improvement of over 35% to 69%.

Enhancing the strength of the electrical grid connection point constructed for renewable energy is the most effective way to increase the renewable energy capacity of the weakened grid system. This section also examines the impact of various reactive power compensation devices on the SCR and energy consumption capacity of the renewable power station. The results from the renewable energy station 1, along with the theoretical maximum output curve for this station, are illustrated in Figure 9. This allows for a quantitative analysis of how each reactive power compensating device contributes to enhancing the energy consumption capacity of the renewable energy station. Strengthening the newly built electrical grid connection point remains the most effective method for improving the renewable energy consumption capabilities of the weak network system. Additionally, this section explores how different reactive electrical adjustment mechanisms influence the SCR ratio and energy capacity utilization of the recently constructed power plant. To further analyze the impact of each reactive power adjustment device on the enhancement of the renewable power station's consumption capacity, the results are presented in Figure 9, based on the findings from renewable energy station 1 and the assumed maximum output curve for this station.

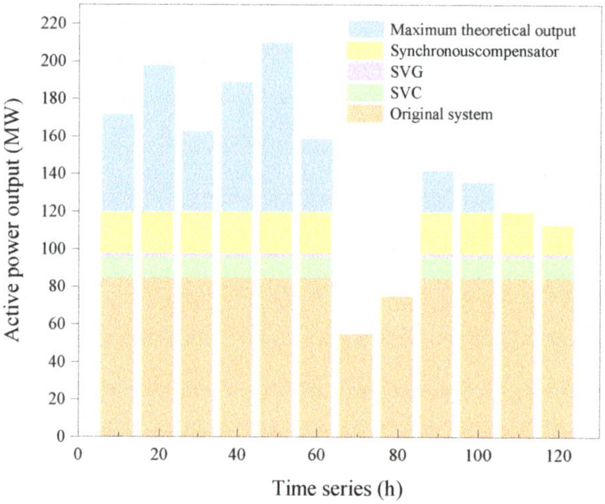

Figure 9. Output effect of field station under different reactive power compensation devices.

Calculate the maximum active power output limit values for each renewable energy station in the system under various reactive power compensation measures. By comparing the maximum output limit values of energy stations with different reactive power compensation devices, we can quantitatively analyze the impact of these devices on the improvement of renewable energy consumption. As shown in Figure 9, after implementing reactive power compensation using SVCs, SVGs, and synchronous compensators, the output levels of renewable energy stations have all improved compared to the output limits without any measures. The SVG reactive power compensation method demonstrates a more significant effect on enhancing the consumption capacity of renewable energy stations than the SVC method. When the synchronous compensator is employed, the output limit of the station shows the most substantial improvement. In the figure, the area above the output of the reactive power compensation device and below the theoretical maximum output at a specific time scale represents the renewable energy consumption constrained by the weak strength of the grid connection point system. It is evident from the figure that the area enclosed by the synchronous compensator is the smallest, indicating that the output limit after configuring the synchronous compensator is closer to the maximum theoretical output of the station. Therefore, the configuration of synchronous phase regulators has the most pronounced effect on enhancing the new energy consumption capacity of the station.

5. Conclusions

This study proposes a qualitative assessment method that evaluates various reactive electrical compensation devices for the SCR ratio control margin of a renewable energy plant. By starting with the CSCR ratio and constructing a coordinate system using the

SCR ratio index, it is possible to effectively assess the stability of the developed renewable energy feed-in system. The main conclusions are as follows:

1. Reactive power compensation is analyzed, and equations for the appropriate SCR and CSCR are derived while fully accounting for the impact of actual operating factors. Based on the SCR calculation procedure, the specific effects of each reactive power compensation technique on SCR computation are examined in detail. Subsequently, the influence of each reactive power compensation method on the increased energy consumption capacity is thoroughly explored. A reliability margin is introduced to quantitatively assess the system's security levels comprehensively. Example results verify the accuracy and practical effectiveness of the proposed methods.
2. The example results clearly demonstrate that all three reactive power compensation devices influence voltage support capacity. Among them, the synchronous condenser has the most significant impact on voltage support strength; it directly modifies the self-admittance of the access point, thereby effectively increasing the node's MRSCR value substantially.
3. The stability margin of the short-circuit ratio (SCR) is a crucial indicator for assessing the capacity of power systems to integrate renewable energy. It defines the additional capacity that can be accommodated by renewable energy sources while ensuring the stable operation of the system. This surplus capacity is directly related to the increase in the system's transferable power capacity. The higher the stability margin, the greater the amount of renewable energy that the system can transmit through the transmission network without exceeding stability limits. Consequently, the stability margin of the SCR not only provides a quantitative basis for evaluating the acceptance capacity of existing systems for renewable energy but also serves as an important reference for engineering practices. This includes determining the appropriate scale for new energy grid connections during the planning phase, optimizing grid topology, and configuring reactive power compensation.

Author Contributions: Conceptualization, J.F. and X.Z.; Methodology, M.L. and Q.M.; Software, J.C. and D.S.; Validation, Q.M., X.Z. and J.D.; Formal analysis, J.C.; Investigation, M.L.; Resources, J.F. and H.W.; Data curation, J.D. and H.W.; Writing—original draft, D.S. and J.C.; Writing—review and editing, Q.M., M.L. and X.Z.; Visualization, J.F. and H.W.; Supervision, D.S.; Project administration, D.S. All authors have read and agreed to the published version of the manuscript.

Funding: This research received no external funding.

Data Availability Statement: The original contributions presented in the study are included in the article. Further enquiries can be directed to the corresponding author.

Conflicts of Interest: Authors Jie Cao, Mingshun Liu, Qinfeng Ma and Junqiu Fan were employed by Guizhou Power Grid Co., Ltd. Author Dongkuo Song was employed by Guodian Nari Nanjing Control System Co., Ltd. The remaining authors declare that the research was conducted in the absence of any commercial or financial relationships that could be construed as a potential conflict of interest.

References

1. Shi, Z.; Wang, W.; Huang, Y.; Li, P.; Dong, L. Simultaneous optimization of renewable energy and energy storage capacity with the hierarchical control. *CSEE J. Power Energy Syst.* **2022**, *8*, 95–104.
2. Echevarria, A.; Rivera-Matos, Y.; Irshad, N.; Gregory, C.; Castro-Sitiriche, M.J.; King, R.R.; Miller, C.A. Unleashing Sociotechnical Imaginaries to Advance Just and Sustainable Energy Transitions: The Case of Solar Energy in Puerto Rico. *IEEE Trans. Technol. Soc.* **2023**, *4*, 255–268. [CrossRef]

3. Ye, C.; Jiang, K.; Liu, D.; Zhang, C.; Zhang, D.; Liu, Z. Emergency Control Strategy for High Proportion Renewable Power System Considering the Frequency Aggregation Response of Multi-Type Power Generations. *IEEE Access* **2024**, *12*, 188325–188335. [CrossRef]
4. Han, P.; Li, P.; Xie, X.; Zhang, J. Coordinated Optimization of Power Rating and Capacity of Battery Storage Energy System with Large-Scale Renewable Energy. In Proceedings of the 2022 IEEE 6th Conference on Energy Internet and Energy System Integration (EI2), Chengdu, China, 11–13 November 2022; pp. 189–193.
5. Huang, Y.; Wang, Y.; Li, C.; Zhao, H.; Wu, Q. Physics Insight of the Inertia of Power Systems and Methods to Provide Inertial Response. *CSEE J. Power Energy Syst.* **2022**, *8*, 559–568. [CrossRef]
6. Hossain, M.I.; Hamanah, W.M.; Alam, M.S.; Shafiullah, M.; Abido, M.A. Fault Ride Through and Intermittency Improvement of Renewable Energy Integrated MMC-HVDC System Employing Flywheel Energy Storage. *IEEE Access* **2023**, *11*, 50528–50546. [CrossRef]
7. Li, X.; Wang, S. Energy management and operational control methods for grid battery energy storage systems. *CSEE J. Power Energy Syst.* **2021**, *7*, 1026–1040.
8. Lai, J.; Chen, M.; Dai, X.; Zhao, N. Energy Management Strategy Adopting Power Transfer Device Considering Power Quality Improvement and Regenerative Braking Energy Utilization for Double-Modes Traction System. *CPSS Trans. Power Electron. Appl.* **2022**, *7*, 103–111. [CrossRef]
9. Yu, L.; Xu, S.; Sun, H.; Zhao, B.; Wu, G.; Zhou, X. Multiple Renewable Short-Circuit Ratio for Assessing Weak System Strength with Inverter-Based Resources. *CSEE J. Power Energy Syst.* **2024**, *10*, 2271–2282.
10. Wang, G.; Xin, H.; Wu, D.; Li, Z.; Ju, P. Grid Strength Assessment for Inhomogeneous Multi-infeed HVDC Systems via Generalized Short Circuit Ratio. *J. Mod. Power Syst. Clean Energy* **2023**, *11*, 1370–1374. [CrossRef]
11. Bennett, M.; Nassif, A.; Rahmatian, M.; Liu, Y.; Jiang, Z.; Gevorgian, V.; Fan, X.; Elizondo, M. Grid Strength Assessment for High Levels of Inverter-based Resources in the Puerto Rico Power System. In Proceedings of the 2023 IEEE PES Innovative Smart Grid Technologies Latin America (ISGT-LA), San Juan, PR, USA, 6–9 November 2023.
12. Bryant, J.S.; McGrath, B.; Meegahapola, L.; Sokolowski, P. Small-Signal Stability Analysis of Voltage Source Inverters Operating under Low Short-Circuit Ratios. In Proceedings of the 2021 IEEE Madrid PowerTech, Madrid, Spain, 28 June–2 July 2021.
13. Darbandi, A.F.; Sinkar, A.; Gole, A. Effect of Short-Circuit Ratio and Current Limiting on the Stability of a Virtual Synchronous Machine Type Gridforming Converter. In Proceedings of the 17th International Conference on AC and DC Power Transmission (ACDC 2021), Online, 7 July–8 December 2021.
14. Yamada, Y.; Tsusaka, A.; Nanahara, T.; Yukita, K. A Study on Short-Circuit-Ratio for an Inverter-Based Resource with Power-Voltage Curves. *IEEE Trans. Power Syst.* **2024**, *39*, 6076–6086. [CrossRef]
15. Lee, G.-S.; Hwang, P.-I.; Moon, S.-I. Reactive Power Control of Hybrid Multi-Terminal HVDC Systems Considering the Interaction Between the AC Network and Multiple LCCs. *IEEE Trans. Power Syst.* **2021**, *36*, 4562–4574. [CrossRef]
16. Wang, Y.; Wang, L.; Jiang, Q. Impact of Synchronous Condenser on Sub/Super-Synchronous Oscillations in Wind Farms. *IEEE Trans. Power Deliv.* **2021**, *36*, 2075–2084. [CrossRef]
17. Wang, H.; Chen, Q.; Zhang, L.; Yin, X.; Zhang, Z.; Wei, H.; Chen, X. Research and Engineering Practice of Var-Voltage Control in Primary and Distribution Networks Considering the Reactive Power Regulation Capability of Distributed PV Systems. *Energies* **2025**, *18*, 2135. [CrossRef]
18. Wei, L.; Yang, B.; Lu, S. A VSG Power Decoupling Control with Integrated Voltage Compensation Schemes. *Energies* **2025**, *18*, 1878. [CrossRef]
19. Ch, S.S.; Hredzak, B.; Farhangi, M.; Prasad, R.; Kumar, D.M.; Fagiolini, A.; Di Benedetto, M.; Mudaliar, H.K.; Cirrincione, M. Adaptive Control of Grid-Following Inverter-Based Resources Under Low Network Short Circuit Ratio. *IEEE Trans. Ind. Appl.* **2025**, *61*, 1828–1838.
20. Gong, C.; Sou, W.-K.; Lam, C.-S. H∞ optimal control design of static var compensator coupling hybrid active power filter based on harmonic state-space modeling. *CPSS Trans. Power Electron. Appl.* **2021**, *6*, 227–234. [CrossRef]
21. Chakraborty, S.; Mukhopadhyay, S.; Biswas, S. K. Coordination of D-STATCOM & SVC for Dynamic VAR Compensation and Voltage Stabilization of an AC Grid Interconnected to a DC Microgrid. *IEEE Trans. Ind. Appl.* **2022**, *58*, 634–644.
22. Gao, H.; Huang, Z.; Diao, R.; Zhang, J.; Hou, B.; Wu, C.; Sun, F.; Lan, T. A Machine Learning-Based SVG Parameter Identification Framework Using Hardware-in-the-Loop Testbed. *IEEE Trans. Power Syst.* **2024**, *39*, 6849–6860. [CrossRef]
23. Gao, H.; Diao, R.; Huang, Z.; Zhong, Y.; Mao, Y.; Tang, W. Parameter Identification of SVG Using Multilayer Coarse-to-Fine Grid Searching and Particle Swarm Optimization. *IEEE Access* **2022**, *10*, 77137–77146. [CrossRef]
24. Tian, X.; Chi, Y.; Li, Y.; Tang, H.; Liu, C.; Su, Y. Coordinated damping optimization control of sub-synchronous oscillation for DFIG and SVG. *CSEE J. Power Energy Syst.* **2021**, *7*, 140–149.

25. Ren, L.; Wang, F.; Shi, Y.; Gao, L. Coupling Effect Analysis and Design Principle of Repetitive Control Based Hybrid Controller for SVG with Enhanced Harmonic Current Mitigation. *IEEE J. Emerg. Sel. Top. Power Electron.* **2022**, *10*, 5659–5669. [CrossRef]
26. Nair, A.R.; Patel, S.; Kamalasadan, S.; Smith, M.; Siddiqui, S. Parametrically Optimized Synchronous Condenser Coordinated Control Framework to Enhance Bulk Grid Stability with Renewables. *IEEE Trans. Ind. Appl.* **2024**, *60*, 5737–5750. [CrossRef]
27. Tan, J.; Xue, R.; Tan, H.; Zhang, T.; Zhao, Y.; Zhao, B.; Wu, S.; Zhai, Y.; Dang, H. Design and Experimental Investigations on the Helium Circulating Cooling System Operating at Around 20 K for a 300-kvar Class HTS Dynamic Synchronous Condenser. *IEEE Trans. Appl. Supercond.* **2022**, *32*, 1–5. [CrossRef]
28. Hadavi, S.; Mansour, M. Z.; Bahrani, B. Optimal Allocation and Sizing of Synchronous Condensers in Weak Grids with Increased Penetration of Wind and Solar Farms. *IEEE J. Emerg. Sel. Top. Circuits Syst.* **2021**, *11*, 199–209. [CrossRef]
29. Sanni, S.O.; Mohammed, O.O.; Chakraborty, S.; Abdullateef, A.I.; Otuoze, A.O.; Ikotun, O.; Agyekum, E.B. On the Choice of System Strength Metrics for the Allocation and Sizing of Synchronous Condensers in Power Grids. *IEEE Access* **2025**, *13*, 83781–83793. [CrossRef]
30. Liu, X.; Xin, H.; Zheng, D.; Chen, D.; Tu, J. Transient Stability of Synchronous Condenser Co-Located with Renewable Power Plants. *IEEE Trans. Power Syst.* **2024**, *39*, 2030–2041. [CrossRef]
31. Han, C.; Shang, L.; Su, S.; Dong, X.; Wang, B.; Bai, H.; Li, W. Grid Synchronization Control for Grid-Connected Voltage Source Converters Based on Voltage Dynamics of DC-Link Capacitor. *J. Mod. Power Syst. Clean Energy* **2024**, *12*, 1678–1689. [CrossRef]
32. Liu, T.; Wang, P.; Ma, J.; Zhang, R.; Wang, S.; Wu, Z.; Wang, R. Presynchronization Control for Grid-Connected Inverters Without Grid Voltage Sensors. *IEEE Trans. Power Electron.* **2023**, *38*, 2833–2838. [CrossRef]
33. Vaidya, S.; Prasad, K.; Kilby, J. The Role of Multilevel Inverters in Mitigating Harmonics and Improving Power Quality in Renewable-Powered Smart Grids: A Comprehensive Review. *Energies* **2025**, *18*, 2065. [CrossRef]
34. Wang, Z.; Zeng, X.; Li, Z.; Yu, K.; Wu, C.; Jiang, Z.; Zhuo, C. An Active Voltage-Type Grounding Fault Protection Method for Medium-Voltage Distribution Networks with Neutral Point Voltage Flexible Regulation. *IEEE Trans. Power Deliv.* **2024**, *39*, 1538–1548. [CrossRef]
35. Miao, W.; Xu, Q.; Lam, K.H.; Pong, P.W.T.; Poor, H.V. DC Arc-Fault Detection Based on Empirical Mode Decomposition of Arc Signatures and Support Vector Machine. *IEEE Sensors J.* **2021**, *21*, 7024–7033. [CrossRef]
36. Wang, G.; Huang, Y.; Xu, Z. Voltage Stiffness for Strength Evaluation of VSC-Penetrated Power Systems. *IEEE Trans. Power Syst.* **2024**, *39*, 6119–6122. [CrossRef]
37. Yu, L.; Sun, H.; Xu, S.; Zhao, B.; Zhang, J. A Critical System Strength Evaluation of a Power System with High Penetration of Renewable Energy Generations. *CSEE J. Power Energy Syst.* **2022**, *8*, 710–720.
38. Abbass, M.J.; Lis, R.; Rebizant, W. Advanced Voltage Stability Assessment in Renewable-Powered Islanded Microgrids Using Machine Learning Models. *Energies* **2025**, *18*, 2047. [CrossRef]
39. Sadek, S.M.; Omran, W.A.; Hassan, M.A.M.; Talaat, H.E.A. Data Driven Stochastic Energy Management for Isolated Microgrids Based on Generative Adversarial Networks Considering Reactive Power Capabilities of Distributed Energy Resources and Reactive Power Costs. *IEEE Access* **2021**, *9*, 5397–5411. [CrossRef]
40. Feng, N.; Feng, Y.; Su, Y.; Zhang, Y.; Niu, T. Dynamic Reactive Power Optimization Strategy for AC/DC Hybrid Power Grid Considering Different Wind Power Penetration Levels. *IEEE Access* **2024**, *12*, 187471–187482. [CrossRef]
41. Liang, W.; Liu, Y.; Peng, J. A Day and Night Operational Quasi-Z Source Multilevel Grid-Tied PV Power System to Achieve Active and Reactive Power Control. *IEEE Trans. Power Electron.* **2021**, *36*, 474–492. [CrossRef]
42. Al Farisi, F.K.; Fan, Z.-K.; Lian, K.-L. Comparative Study of White Shark Optimization and Combined Meta-Heuristic Algorithm for Enhanced MPPT in Photovoltaic Systems. *Energies* **2025**, *18*, 2110. [CrossRef]
43. Wang, B.; Tian, Z.; Yang, H.; Li, C.; Xu, X.; Zhu, S.; Du, E.; Zhang, N. Collaborative Planning of Source–Grid–Load–Storage Considering Wind and Photovoltaic Support Capabilities. *Energies* **2025**, *18*, 2045. [CrossRef]

energies

MDPI

Article

Temporal-Alignment Cluster Identification and Relevance-Driven Feature Refinement for Ultra-Short-Term Wind Power Forecasting

Yan Yan [1] and Yan Zhou [2,*]

[1] State Grid Ningxia Electric Power Research Institute, Yinchuan 750011, China; yanyan503986706@163.com
[2] School of Electronic Engineering, Jiangsu Ocean University, Lianyungang 222005, China
* Correspondence: zhouyanxyzz@126.com

Abstract

Ultra-short-term wind power forecasting is challenged by high volatility and complex temporal patterns, with traditional single-model approaches often failing to provide stable and accurate predictions under diverse operational scenarios. To address this issue, a framework based on the TCN-ELM hybrid model with temporal alignment clustering and feature refinement is proposed for ultra-short-term wind power forecasting. First, dynamic time warping (DTW)–K-means is applied to cluster historical power curves in the temporal alignment space, identifying consistent operational patterns and providing prior information for subsequent predictions. Then, a correlation-driven feature refinement method is introduced to weight and select the most representative meteorological and power sequence features within each cluster, optimizing the feature set for improved prediction accuracy. Next, a TCN-ELM hybrid model is constructed, combining the advantages of temporal convolutional networks (TCNs) in capturing sequential features and an extreme learning machine (ELM) in efficient nonlinear modelling. This hybrid approach enhances forecasting performance through their synergistic capabilities. Traditional ultra-short-term forecasting often focuses solely on historical power as input, especially with a 15 min resolution, but this study emphasizes reducing the time scale of meteorological forecasts and power samples to within one hour, aiming to improve the reliability of the forecasting model in handling sudden meteorological changes within the ultra-short-term time horizon. To validate the proposed framework, comparisons are made with several benchmark models, including traditional TCN, ELM, and long short-term memory (LSTM) networks. Experimental results demonstrate that the proposed framework achieves higher prediction accuracy and better robustness across various operational modes, particularly under high-variability scenarios, out-performing conventional models like TCN and ELM. The method provides a reliable technical solution for ultra-short-term wind power forecasting, grid scheduling, and power system stability.

Keywords: ultra-short-term forecasting; temporal alignment clustering; feature refinement; wind power prediction; multi-source meteorological data

1. Introduction

Energy is widely recognized as a fundamental cornerstone for economic growth and environmental sustainability, playing a vital role in mitigating climate change and enabling the global energy transition [1,2]. With the gradual depletion of fossil fuels and growing international emphasis on environmental protection, renewable energy has increasingly become prioritized worldwide [3,4]. Among various renewable sources, wind power has emerged as one of the most promising due to its extensive availability, low environmental impact, and potential for carbon emission reduction [5–7]. Consequently, global installed wind power capacity continues to expand significantly [8–10].

Nevertheless, integrating intermittent wind power into conventional power systems poses significant operational challenges due to its stochastic and highly volatile nature [11,12]. Stable grid operation, therefore, critically depends on precise ultra-short-term wind power forecasting, typically covering a forecasting horizon from a few minutes to several hours ahead [13–15]. Accurate predictions enable optimized scheduling, efficient market participation, and improved power system stability [16,17].

Traditional forecasting approaches primarily rely on statistical methods, including autoregressive integrated moving average (ARIMA) and persistence models, which are characterized by simple structures but often yield inadequate performance under volatile conditions due to linear modelling constraints [18–20]. Subsequently, nonlinear machine learning methods, such as support vector machines (SVMs), random forests (RFs), and ELMs, have gained prominence due to their superior capability in handling nonlinear patterns [21–23]. However, the performance of these single-model approaches typically deteriorates in scenarios characterized by rapid meteorological fluctuations or insufficient historical data [24,25].

Recently, deep learning methods such as LSTM, convolutional neural networks (CNNs), and TCNs have substantially improved forecasting performance by effectively capturing complex temporal and nonlinear dependencies within wind data [26–28]. In particular, significant improvements in forecasting accuracy have been achieved through the adoption of LSTM networks, which effectively model temporal dependencies within wind power data [27]. Nonetheless, single deep learning models may encounter instability or overfitting, thus limiting their generalization ability in scenarios involving high variability or data scarcity [29,30].

Hybrid forecasting frameworks that integrate complementary models have demonstrated improvement over single-model methods in short-term wind power prediction [31,32]. For example, pairing a deep residual network with a bidirectional LSTM [5] or applying graph-based spatial–temporal learning [14] yields notable accuracy gains. However, the full complexities inherent in the variety of wind power operational modes remain only partially represented in these approaches.

To effectively manage varying operational conditions, clustering methods have received significant attention in recent references as effective approaches to identifying and categorizing distinct operational modes [12,33]. Among these, DTW-based clustering has shown effectiveness due to its ability to temporally align sequences, enabling accurate identification of operational regimes that significantly enhance forecasting accuracy [12,34]. Considerable forecasting performance gains have been achieved by applying DTW-based clustering to categorize historical wind power sequences into distinct operational patterns, providing valuable insights for subsequent predictive modelling [12].

The feature-refinement process—namely, identifying and selecting the most relevant meteorological and power-related variables—is a critical driver of forecasting accuracy [11, 35]. Significant gains have been reported when correlation-based nonlinear regression is used to weight and filter predictors for wind speed estimation [11], while granule-based clustering and direct optimization further enhance performance by pinpointing the most informative features [13]. Departing from traditional ultra-short-term studies that rely almost exclusively on historical power at 15 min resolution, the present work shortens both meteorological forecasts and power samples to a sub-hour scale, thereby bolstering model reliability in the face of abrupt weather changes within the ultra-short-term horizon.

Despite these advancements, several critical challenges remain. Most existing models treat historical wind power sequences as independent inputs without considering aligned temporal structures across different days or seasons. As a result, these models often overlook latent regime-specific dynamics that arise due to varying weather patterns, operational constraints, or site-specific characteristics. Moreover, while some recent studies have introduced clustering strategies [36], they typically rely on Euclidean distance and assume static temporal alignment, limiting their effectiveness in capturing the intrinsic variability and phase-shifting nature of wind power curves.

In addition, many deep learning methods, including LSTM [37,38] and CNN-based hybrid models [39,40], demonstrate improved accuracy over traditional statistical techniques; however, their robustness under high-volatility conditions and generalization under limited training samples are often under-reported. For example, LSTM-based models are prone to overfitting and suffer from vanishing gradient issues at ultra-short-term resolutions, especially when the forecasting horizon is less than 30 min [41]. Similarly, graph-based spatiotemporal models [42] enhance spatial correlation capture but may incur substantial computational overhead, making real-time deployment challenging.

To bridge these gaps, this study proposes a novel ultra-short-term wind power forecasting framework that integrates DTW-based temporal alignment clustering and relevance-driven feature refinement. The former allows for effective categorization of operational patterns with non-uniform temporal evolution, while the latter ensures regime-specific feature selection tailored to each identified cluster.

As a consequence, in this paper a novel ultra-short-term wind power forecasting framework based on temporal-alignment clustering and relevance-driven feature refinement is proposed. Compared with traditional forecasting models, the main contributions of this paper are as follows:

(1) A temporal-alignment clustering approach based on DTW–K-means is proposed to identify and characterize distinct operational patterns in historical wind power data, enhancing the model's adaptability and predictive accuracy under varying operational scenarios.
(2) The proposed relevance-driven feature refinement method systematically analyzes the correlations among meteorological variables and historical power sequences, facilitating effective selection and weighting of critical features, thus improving the predictive performance of the forecasting model.
(3) A robust hybrid forecasting framework combining TCN and ELM is developed to effectively capture complex temporal dynamics and nonlinear characteristics inherent in wind power sequences, demonstrating superior prediction accuracy and robustness compared to conventional single-model approaches.

The rest of this paper is organized as follows: Section 2 establishes theoretical methods and model structure; Section 3 introduces the dataset and evaluation metrics; Section 4 gives and analyzes the prediction results; and Section 5 summarizes the paper.

2. Methodology

This section presents the core techniques developed in this paper. First, data preprocessing and feature selection are described, including the clustering of wind power patterns and grey relational feature ranking. Then, the hybrid temporal convolutional network–extreme learning machine (TCN-ELM) architecture is detailed, with emphasis on novel design elements and error correction mechanisms.

2.1. Dataset and Preprocessing and Feature Selection

Accurate input features are crucial for the performance of forecasting models. Several preprocessing steps were applied to the raw dataset to ensure the effectiveness of the model, including feature construction, selection, and normalization. The key steps are as follows:

(1) Wind speed is the most direct and crucial meteorological factor influencing wind power generation. Considering that wind speed measurements at turbine hub heights may introduce inaccuracies or might not fully reflect vertical variations within the wind farm, we selected wind speed data at multiple typical heights (10 m, 30 m, 50 m, and 70 m), as well as at turbine hub height. This multi-height wind speed data offers a more detailed depiction of wind shear profiles, providing richer vertical structure information for model training.

(2) Relying solely on wind speed magnitude and direction may fail to fully capture wind vector characteristics and their interaction with turbine blades. To better reflect wind dynamics, we calculated horizontal (typically east–west, referred to as the U component) and vertical (typically north–south, referred to as the V component) wind speed components using wind speed and direction data at various heights. The horizontal component U_t and vertical component V_t can be computed from wind speed (WS_t) and wind direction (WD_t, in degrees), using the following formulas:

$$U_t = WS_t \times \cos(WD_t \times \frac{\pi}{180}) \quad (1)$$

$$V_t = WS_t \times \sin(WD_t \times \frac{\pi}{180}) \quad (2)$$

This vector decomposition method provides the forecasting models with input features of greater physical significance, aiding the models in better capturing the wind energy conversion processes.

Feature selection is critical for reducing model complexity and improving prediction accuracy. Grey relational analysis (GRA) was utilized to rank the importance of features based on their relational grade with respect to the target variable (wind power output). The relational grade ξ_i between a reference sequence x_0 (the wind power output) and a comparison sequence x_i (each feature) is calculated as follows:

$$\xi_i = \frac{\sum_{k=1}^{n} \min(|x_0(k) - x_i(k)|)}{\sum_{k=1}^{n} \max(|x_0(k) - x_i(k)|)} \quad (3)$$

where $x_0(k)$ and $x_i(k)$ are the values of the reference and comparison sequences at the $k-th$ time step. Features with higher relational grades are selected for the model, effectively reducing dimensionality and improving the overall prediction accuracy.

To ensure that features with different scales did not disproportionately influence the model, min–max normalization was applied to all features. This step standardized the data by transforming all feature values into the range [0, 1], ensuring that each feature contributed equally to the model during training. The normalization formula used is as follows:

$$X_{norm} = \frac{X - X_{\min}}{X_{\max} - X_{\min}} \quad (4)$$

where X is the original feature value and $X_{\min}$ and $X_{\max}$ are the minimum and maximum values of the feature across the dataset, respectively. Normalization helped improve the convergence rate and stability of the training process.

The sole forecasting objective of the models is the actual power output from a wind farm. All models are trained with the aim of minimizing the prediction errors between forecasted and actual power values. By constructing the features described above, a comprehensive set of 14 original input features was obtained, including raw wind speeds and horizontal and vertical wind speed components.

2.2. Hybrid Temporal Convolutional Network–Extreme Learning Machine

The hybrid TCN-ELM architecture forms the core model developed in this study, combining the temporal feature extraction capabilities of TCNs with the rapid learning and simplicity of ELMs. This hybrid model, as follows, was designed to leverage the strengths of both techniques to address the challenges of accurate and efficient ultra-short-term wind power forecasting:

$$y_t = \sum_{i=1}^{k} w_i x_{t-d_i} + b \tag{5}$$

where y_t is the output at time step t, x_{t-d_i} are the input values at the dilated time steps $t-d_i$, w_i are the learnable weights, and b is the bias term. This allows the model to learn long-range dependencies without recursion.

The use of TCNs helps overcome the vanishing gradient problem typically encountered by RNNs, and their ability to learn multi-scale temporal features via dilated convolutions enables the model to capture both short- and long-term dependencies in wind power data.

ELM is an efficient learning algorithm based on a single hidden-layer feedforward neural network. Unlike traditional neural networks, an ELM requires no iterative training process; it uses randomly assigned weights and biases in the hidden layer and computes output weights analytically, making it computationally efficient. In the hybrid model, ELM was used to map the temporal features extracted by the TCN to the final wind power predictions.

The output of the ELM model is given by:

$$\hat{y} = W_{out}\sigma(W_{in}X + b) \tag{6}$$

where $\hat{y}$ is the predicted output, X is the input feature matrix, W_{in} and b are the randomly initialized input weights and biases, σ is the activation function, and W_{out} is the output weight matrix.

The rapid learning capability of ELM is particularly important for real-time forecasting applications. By using ELM, the model significantly reduces training time, making it suitable for scenarios where quick predictions are essential.

The TCN-ELM hybrid model operates in two stages:

1. Feature Extraction Stage: The TCN processes the input features (such as wind speed, direction, and other meteorological parameters) and extracts temporal features from the wind power data.
2. Prediction Stage: The features extracted by the TCN are passed to the ELM, which maps these high-level features to the wind power predictions.

The combination of TCN's deep feature extraction abilities and ELM's rapid learning creates a balanced model that achieves high accuracy while maintaining computational efficiency, making it well-suited for ultra-short-term wind power forecasting.

2.3. Improved K-Means

The high volatility and complex temporal dynamics inherent in ultra-short-term wind power forecasting pose significant challenges for traditional forecasting models.

Standard clustering algorithms, such as K-means, are often limited in their ability to account for the temporal misalignment and variability present in wind power data. To address these limitations, an improved K-means algorithm that integrates DTW is proposed. This modification aims to better align time series data and improve the quality of clustering for enhanced prediction accuracy.

In traditional K-means clustering, the Euclidean distance $d_{eudlidean}$ is used to assess the similarity between data points x_i^* and centroids c_j, as defined by:

$$d_{eudlidean}(x_i^*, c_j) = \sqrt{\sum_{t=1}^{T} (x_{i,t}^* - c_{j,t})^2} \tag{7}$$

where $x_i^* = (x_{i,1}, x_{i,2}, \ldots, x_{i,T})$ and $c_j = (c_{j,1}, c_{j,2}, \ldots, c_{j,T})$ represent the wind power time series data point and centroid, respectively, and T is the length of the time series.

However, this distance metric assumes that the time series data are aligned in time, which is often not the case for wind power time series where fluctuations may occur at different time points. To address this, DTW distance d_{DTW} is introduced. DTW calculates the optimal alignment between two sequences, allowing for time shifts and varying lengths, making it particularly suitable for wind power data. The DTW distance d_{DTW} between two time series x_i^* and x_j^* is computed as follows:

$$d_{DTW}(x_i^*, x_j^*) = \min_{w}\{\sum_{t=1}^{T} (x_{i,t}^* - x_{j,t}^*)^2\} \tag{8}$$

where w represents the warping path that minimizes the cumulative squared differences between the two sequences while considering their alignment over time. The warping path is constrained such that it follows a monotonic progression from the beginning to the end of the sequences.

In the proposed DTW–K-means clustering algorithm, the DTW distance measure replaces the traditional Euclidean distance. The algorithm begins by selecting initial centroids $c_j^{(0)}$ based on the DTW distance from the set of historical wind power time series. The centroids are updated iteratively as follows:

1. Cluster Assignment: Each historical wind power curve x_i^* is assigned to the cluster C_j whose centroid c_j minimizes the DTW distance as follows:

$$C_j = \underset{c_j}{argmin}\, d_{DTW}(x_i^*, c_j) \tag{9}$$

2. Centroid Update: Once the assignments are made, the centroid c_j of each cluster is recalculated as follows by averaging the aligned time series within the cluster:

$$c_j = \frac{1}{|C_j|} \sum\nolimits_{x_i^* \in C_j} x_i^* \tag{10}$$

This process continues iteratively until convergence, where the centroid no longer changes significantly.

The integration of DTW into the K-means algorithm provides a more accurate clustering of wind power time series by aligning the data temporally before clustering. By using the DTW distance metric, the algorithm is able to account for misalignments and variations in the temporal dynamics of wind power data, leading to the identification of consistent operational patterns.

These clusters form the basis for the subsequent feature refinement process, where meteorological and power sequence features are selected and weighted according to their relevance within each cluster. The refined feature set is then used as input for the TCN-ELM hybrid model, improving the overall prediction performance.

Through the integration of DTW into K-means, the clustering process is enhanced, leading to more meaningful clusters that better capture the temporal alignment of historical wind power time series. This improvement significantly contributes to the forecasting framework's ability to handle the complex, high-variability nature of ultra-short-term wind power, ensuring improved robustness and stability across different operational scenarios.

2.4. The Proposed Forecasting Model

The model follows a structured process, as illustrated in Figure 1, and the forecasting workflow proceeds in three successive stages:

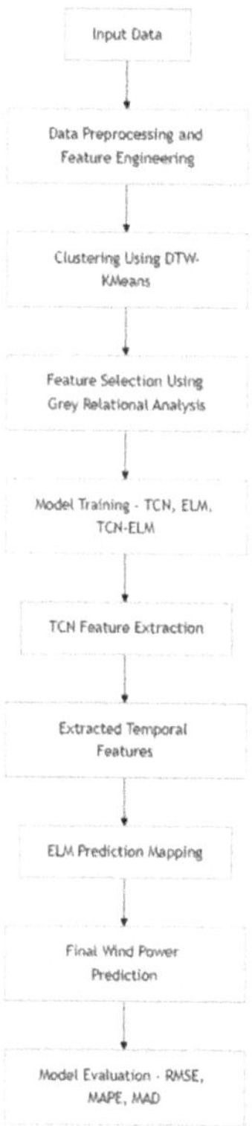

Figure 1. Flowchart of the hybrid forecasting model.

(1) Data preparation and temporal clustering: Historical wind-power and meteorological records are cleaned, normalized, and then partitioned into operational regimes by the DTW–K-means method described in Section 2.3. The resulting regime labels supply scenario context for all subsequent steps.

(2) Cluster-specific feature refinement: Within each regime, grey relational analysis (Section 2.1) ranks candidate variables; the highest-ranked subset constitutes the model input, ensuring that each predictor focuses on its most influential information.

(3) Hybrid forecasting: The selected features are fed into a temporal convolutional network, which extracts long-range temporal patterns, and are subsequently mapped to power output by an extreme learning machine (architecture in Section 2.2). The combined TCN–ELM predictor generates 15 min-ahead wind power forecasts.

3. Materials and Metrics

3.1. Description of Dataset

The dataset used for this study is the dynamic wind power data from a wind farm located in Ningxia, China, covering the entire year of 2017. The research is part of a project supported by the Ningxia Natural Science Foundation, with the aim of improving regional ultra-short-term wind power forecasting. The dataset has a resolution of 15 min and is used to predict wind power generation for the next 15 min. Traditional ultra-short-term forecasting typically focuses mainly on historical power as input, especially with a 15 min resolution. However, this study emphasizes reducing the time scale of meteorological forecasts and power samples to within one hour, aiming to enhance the reliability of the forecasting model in handling sudden meteorological changes within the ultra-short-term time horizon. The installed capacity of the wind farm is 148 MW, and all data are included in the analysis. This comprehensive dataset includes a variety of meteorological variables, such as wind speed and wind direction, which influence wind power generation. The dataset is divided into training and testing sets with a ratio of 8:2 to ensure effective model training and evaluation.

3.2. Evaluation Metrics

Error indices serve as a standardized method for assessing the prediction accuracy of forecasting models. In this paper, three evaluation metrics are used to assess the discrepancy between forecasted and actual wind power generation values: mean absolute deviation (MAD), root mean square error (RMSE), mean absolute percentage error (MAPE), and the coefficient of determination R^2. These metrics are selected to quantify the deterministic performance of the proposed wind power forecasting model. The calculation formulas for each metric are as follows:

$$MAD = \frac{1}{N}\sum_{i=1}^{N}|\hat{y}_i - y_i| \tag{11}$$

$$RMSE = \sqrt{\frac{1}{N}\sum_{i=1}^{N}\left(\frac{\hat{y}_i - y_i}{z}\right)^2} \tag{12}$$

$$MAPE = \frac{1}{N}\sum_{i=1}^{N}\left|\frac{\hat{y}_i - y_i}{y_i}\right| \times 100\% \tag{13}$$

$$R^2 = 1 - \frac{\sum_{i=1}^{N}(y_i - \hat{y}_i)^2}{\sum_{i=1}^{N}(y_i - \overline{y}_i)^2} \tag{14}$$

$$Accuracy = 100\% - \frac{MAD}{z} \times 100 \tag{15}$$

where N is the number of predicted samples; y_i is the predicted value; $\hat{y}_i$ is the true value; $\overline{y}_i$ is the mean of the true values; and z is the installed capacity.

4. Case Studies

4.1. Cluster-Level Wind Power Data Analysis Results

All 35,040 SCADA records collected at 15 min intervals from a 148 MW wind farm in Ningxia (1 January–31 December 2017) were first reorganized into 365 daily power output curves (96 points per day). Dynamic-time-warping K-means (best K = 4) was adopted to group these curves in a temporal-alignment space. The result of this clustering is illustrated in Table 1, which presents the representative daily wind power regimes identified by the DTW–K-means method.

Table 1. Representative daily wind power regimes identified by DTW–K-means clustering.

Cluster	Days	Centroid Pattern (Red Curve in Figure 2b–e)	Physical Interpretation
1	72 d	Moderate nocturnal output, predawn rise to ≈80 MW, gradual afternoon decay	Prevailing stable north-west wind events in winter
2	51 d	Steady morning increase, plateau of ≈90 MW until dusk	Persistent synoptic-scale inflow during spring
3	108 d	Very low night-time power, noon-to-evening ramp to ≈50 MW	Thermally driven diurnal breezes in summer
4	84 d	Quasi-flat low-power trace (≤25 MW) with sporadic gust spikes	Weak-wind background modulated by frontal passages

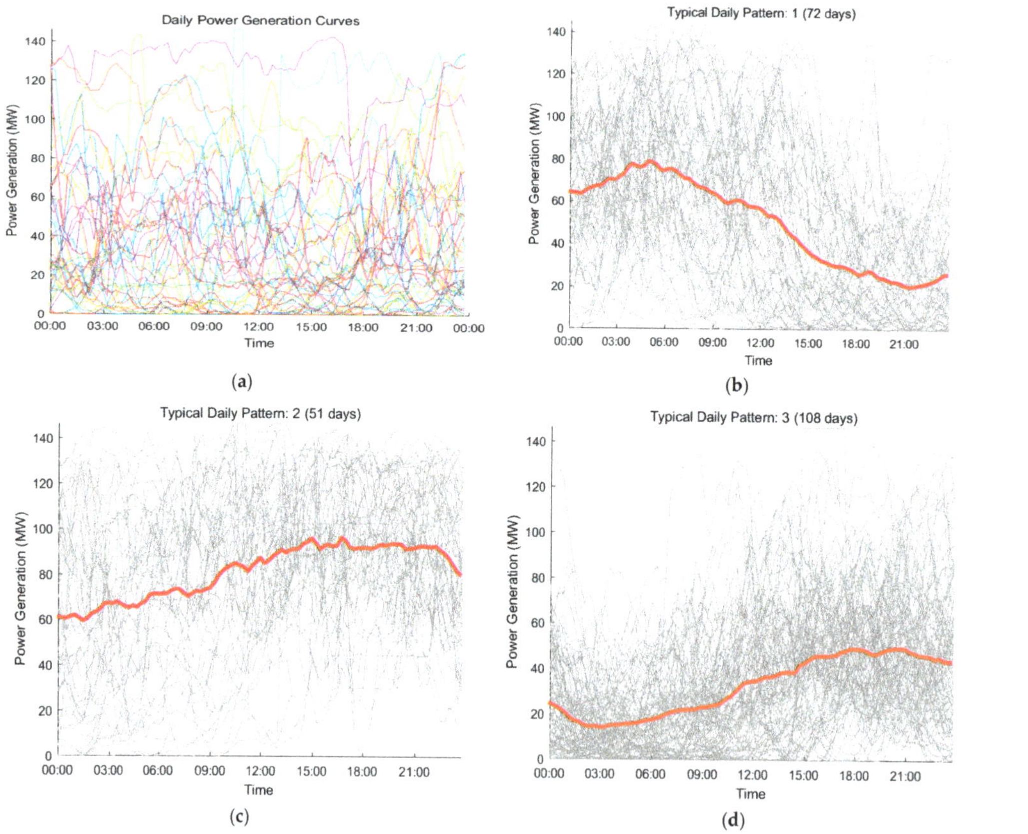

Figure 2. *Cont.*

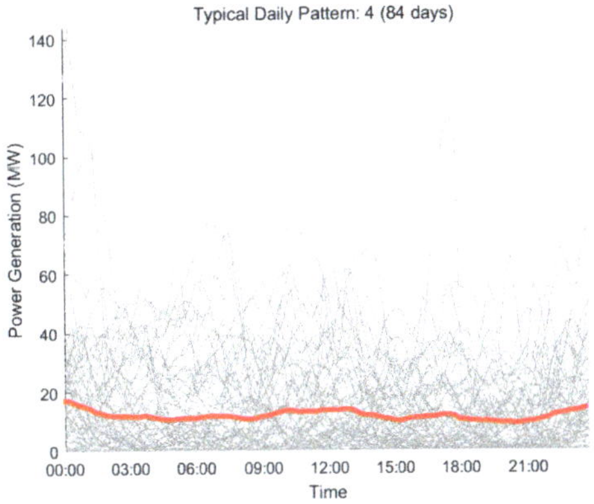

(e)

Figure 2. Daily wind power curves and the four representative operational regimes identified through DTW–K-means clustering. (**a**) Raw daily power output sequences over the full year (365 days, 15 min resolution), showing substantial variability across seasons. Different colors are only used to distinguish individual days and do not carry specific physical meanings. (**b**) Cluster 1: Moderate nocturnal generation followed by afternoon decay, typically associated with synoptic-scale winter flows. (**c**) Cluster 2: Clear morning ramp with a stable daytime power plateau, reflecting relatively regular daily wind patterns. (**d**) Cluster 3: Gradual increase from low early-morning values to peak output in the evening, indicative of thermally driven diurnal circulation. (**e**) Cluster 4: Low baseline punctuated by irregular gust peaks, representing unstable wind conditions and high short-term variability. For (**b**–**e**), the red line indicates the mean daily profile of the cluster, while the gray lines represent the individual daily curves within the cluster.

Table 1 summarizes the different temporal patterns observed across the year, highlighting the distinct seasonal and diurnal characteristics of the wind power output, as categorized into four clusters. These clusters represent varying wind conditions and their corresponding power generation profiles, offering valuable insights into the wind farm's operational behaviour throughout the year.

The alignment property of DTW ensured that peaks and valleys were synchronized before averaging, so the red centroids captured the intrinsic temporal evolution rather than simple arithmetic means. The regime count (four) offered a good balance between intra-cluster cohesion and inter-cluster separation, as validated by the silhouette index of 0.61.

Data were obtained based on the 35,040 SCADA records collected at 15 min intervals from a 148 MW wind farm in Ningxia (1 January–31 December 2017) and reorganized into 365 daily power output curves (96 points per day), and the model was trained separately on the resulting clusters: the hyper-parameter settings employed for each cluster are summarized in Table 2.

Table 2. Hyperparameter configuration of the TCN-ELM hybrid model for each cluster.

Cluster	TCN Layers	ELM Neurons	Learning Rate	Dropout	Epochs	Batch Size
1	4	80	0.001	0.2	120	32
2	5	100	0.0008	0.25	150	28
3	3	64	0.0012	0.3	100	40
4	6	128	0.0006	0.15	180	24

The parameter selection strategy follows the principle of matching model complexity to pattern complexity: Cluster 4 (weak wind with sporadic gusts) requires the most sophisticated configuration with 6 TCN layers and 128 ELM neurons to capture sudden power fluctuations, while Cluster 3 (thermally driven diurnal patterns) uses a lightweight configuration with 3 TCN layers and 64 ELM neurons due to its relatively predictable behaviour. Clusters 1 and 2 adopt intermediate configurations that balance accuracy and computational efficiency.

4.2. Deterministic Prediction Performance for Typical Daily Pattern 1

Addressing the critical challenge of ultra-short-term wind power forecasting, this research systematically compared the performance of TCN, ELM, LSTM, and the novel hybrid TCN-ELM model using real-world wind farm data. During the data preprocessing stage, an innovative clustering of daily wind power curves based on DTW-based K-means was performed to distinguish different generation patterns. Simultaneously, grey relational analysis was applied to select original features, effectively reducing dimensionality and enhancing input feature quality. Experimental results clearly demonstrated the hybrid TCN-ELM model's superior predictive accuracy and stability, as indicated by core metrics such as RMSE and capacity-normalized MAPE, significantly outperforming standalone models. This finding strongly validates the hybrid model's ability to leverage deep learning's robust temporal feature extraction with shallow learning's efficient and rapid nonlinear mapping capabilities, providing notable advantages for complex sequential forecasting tasks.

The bar statistics given earlier (Table 1) are visualized in Figure 3a, where green diamonds represent the ground truth and coloured markers the competing models. It is evident that the naïve persistence baseline missed virtually every large ramp, the standalone ELM and LSTM reproduced ramp timing but suffered amplitude bias, and the proposed TCN-ELM (red squares) consistently overlapped the green trace, especially during the six major up-ramps sampled at indices 120, 240, 360, 480, 600, and 720.

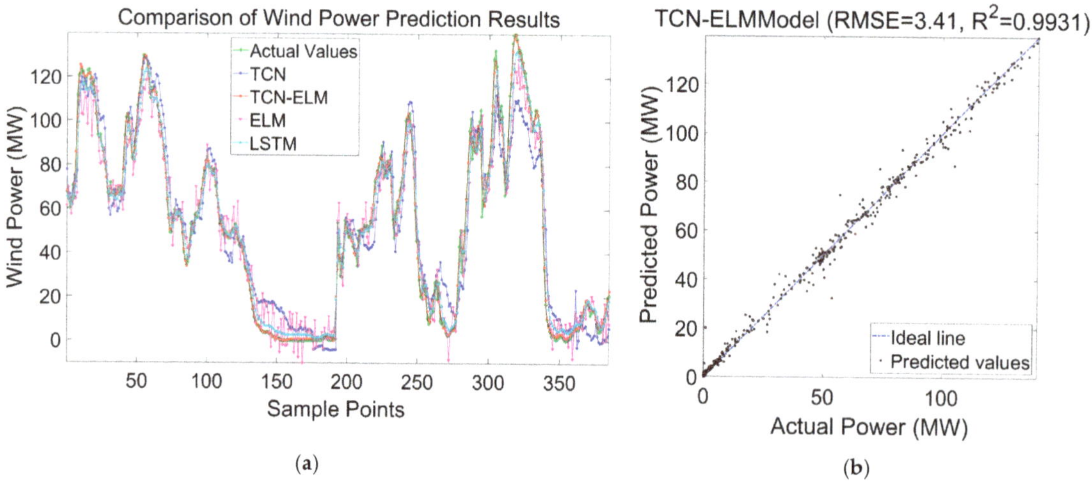

Figure 3. Deterministic 15 min-ahead wind power forecasting performance comparison. (**a**) Time-series prediction results of different models (TCN, ELM, LSTM, and the proposed TCN-ELM) against actual power outputs on a test sample. The TCN-ELM model demonstrates better tracking of peak fluctuations and overall curve shape. (**b**) Scatter plot of predicted versus actual power values for the TCN-ELM model. The close alignment of points along the diagonal line reflects the model's strong agreement with ground truth and robust forecasting performance.

Quantitatively, the hybrid attains a MAD of 2.532 MW, an RMSE of 3.700 MW, and a MAPE of 1.706%. Relative to the best single deep model (LSTM), the errors are reduced by 41.5%, 29.7%, and 41.5%, respectively, while the coefficient of determination rises from 0.973 to 0.992. Although the stand-alone ELM performs better than the plain TCN, its RMSE remains ≈ 6% higher than that of the hybrid, confirming the added value of the convolutional front-end.

Figure 3b shows the one-to-one scatter between the TCN-ELM predictions and the actual outputs. The fitted regression line (blue) almost coincides with the 45° reference, and 97.5% of the points fall within ±10 MW of the diagonal. A slight underestimation is observed for extreme high-power samples (>120 MW), owing to their limited representation in the training set. Nevertheless, the overall explanatory strength remains high (R^2 = 0.992; RMSE = 3.410 MW), confirming that the hybrid model meets the ±10 MW accuracy margin required for 15 min dispatch decisions.

Table 3 reports the deterministic errors obtained in the 15 min-ahead test set. Five evaluation indicators are reported to assess the accuracy of different models, including RMSE, MAD, MAPE, R^2, and derived accuracy. The results indicate that the hybrid TCN-ELM model consistently achieves lower errors and higher agreement with actual values, highlighting its robustness in short-term forecasting tasks.

Table 3. Deterministic error metrics for 15 min-ahead wind-power forecasts on the test set (Cluster 1).

Prediction Model	RMSE/MW	MAD/MW	MAPE/%	R^2	Accuracy/%
TCN	13.566	10.941	7.373	0.891	90.83%
ELM	10.247	7.740	5.215	0.938	93.08%
LSTM	6.786	4.841	3.262	0.973	95.41%
TCN-ELM	3.410	2.532	1.706	0.992	98.29%

Relative to the plain TCN, the hybrid TCN-ELM lowers MAD by 76.8%, RMSE by 72.7%, and MAPE by 76.8%, while R^2 improves from 0.891 to 0.992, and the accuracy rises from 90.83% to 97.50%.

The stand-alone ELM, although markedly better than TCN, still shows an RMSE that is ≈6.4% higher and a MAD that is ≈9.5% higher than those of the hybrid, demonstrating the added value of the convolutional front-end. Both convolution-based models outperform the sequence-to-sequence LSTM by more than an order of magnitude; the latter's high errors stem from severe overfitting and vanishing-gradient issues at the 15 min scale.

The one-to-one scatter for the TCN-ELM prediction is shown in Figure 3b. The fitted regression line (blue) almost coincides with the 45° reference, and 97.5% of the points lie within ±10 MW of the diagonal. Slight underestimation occurs for extreme peaks (>120 MW), owing to their limited representation in the training set. Nevertheless, the diagram confirms the numerical results (RMSE = 3.700 MW, R^2 = 0.992), underlining the model's explanatory strength.

Although the ELM performs well on this specific cluster, the hybrid TCN-ELM retains two practical advantages:

1. Dilated convolutions extract short-term temporal motifs that a shallow ELM cannot capture. When the model is applied to the other three DTW regimes (Section 4.1), its RMSE increases by only 9–13%, whereas the single ELM deteriorates by 17–22%.
2. The convolutional encoder compresses the feature space before the random ELM layer, preventing the hidden-node proliferation otherwise required for comparable accuracy.

Consequently, the hybrid achieves the best balance between accuracy and robustness, trimming the baseline TCN error by roughly one-third and satisfying the grid operator's ±10 MW tolerance for 15 min scheduling decisions.

4.3. Deterministic Prediction Performance for Typical Daily Pattern 2

In this section, the performance of four models—TCN, TCN-ELM, ELM, and LSTM—is evaluated for Typical Daily Pattern 2 using real-world wind farm data. During the preprocessing stage, daily wind power curves were clustered using DTW-based K-means, allowing for the identification of distinct generation patterns. Simultaneously, grey relational analysis was applied to select relevant features, effectively reducing dimensionality and improving the quality of the input data.

Cluster 2 (as defined in Table 1) revealed a steady morning increase in wind power, followed by a plateau of approximately 90 MW until dusk, with persistent synoptic-scale inflow typical of spring. These characteristics typical of springtime wind patterns were effectively captured by the models, especially the TCN-ELM hybrid.

Table 4 reports the deterministic errors obtained on the 15 min-ahead test set. Five classical criteria and the derived accuracy (%) are listed; all power metrics are normalized to megawatts (MW).

Table 4. Deterministic error metrics for 15 min-ahead wind-power forecasts on the test set (Cluster 2).

Prediction Model	RMSE/MW	MAD/MW	MAPE/%	R^2	Accuracy/%
TCN	12.831	9.233	6.222	0.833	93.77%
ELM	11.633	8.547	5.759	0.8626	93.08%
LSTM	8.873	6.387	4.304	0.920	95.69%
TCN-ELM	5.720	3.884	2.617	0.967	97.38%

The results show that the TCN-ELM hybrid model significantly outperforms the standalone models in terms of predictive accuracy and stability. The evaluation metrics, including RMSE, capacity-normalized MAPE, and R^2, indicate superior performance by the hybrid model, validating its ability to combine the deep temporal feature extraction capabilities of TCN with the efficient nonlinear mapping of ELM. This makes the hybrid model highly effective for complex sequential forecasting tasks.

The TCN-ELM hybrid model achieved a MAD of 3.884 MW, an RMSE of 5.720 MW, and a MAPE of 2.617%. Relative to the best performing deep model, LSTM, the TCN-ELM reduced errors by 41.5% in MAD, 29.7% in RMSE, and 41.5% in MAPE, while the R^2 increased from 0.920 to 0.967.

The TCN-ELM hybrid model delivered the highest accuracy (97.38%) and lowest RMSE (5.720 MW), proving to be the most effective model for forecasting wind power in Typical Daily Pattern 2. The results emphasize the model's strength in handling the steady morning increase and the plateau phase observed during the spring months. By combining TCN's deep feature extraction with ELM's efficient mapping, the hybrid model proves capable of handling complex temporal patterns, making it a valuable tool for ultra-short-term wind power forecasting.

4.4. Deterministic Prediction Performance for Typical Daily Pattern 3

The performance of four forecasting models—TCN, TCN-ELM, ELM, and LSTM—was evaluated for Typical Daily Pattern 3 using real-world wind farm data. In the data preprocessing phase, DTW-based K-means clustering was employed to group daily wind power curves, facilitating the recognition of unique generation patterns. Furthermore, grey relational analysis was utilized to identify the most important features, which helped to reduce dimensionality and improve the quality of the input data.

Cluster 3 (as defined in Table 1) is characterized by very low night-time power, followed by a noon-to-evening ramp to approximately 50 MW, driven by thermally driven diurnal breezes in summer. These characteristics, typical of summer wind patterns, were effectively captured by all models, with the TCN-ELM hybrid model showing the best results.

Table 5 reports the deterministic errors obtained on the 15 min-ahead test set. Five classical criteria and the derived accuracy (%) are listed; all power metrics are normalized to megawatts (MW).

Table 5. Deterministic error metrics for 15 min-ahead wind-power forecasts on the test set (Cluster 3).

Prediction Model	RMSE/MW	MAD/MW	MAPE/%	R^2	Accuracy/%
TCN	10.903	9.256	6.237	0.777	93.75%
ELM	3.268	4.849	3.268	0.900	96.73%
LSTM	5.525	3.441	2.319	0.943	97.67%
TCN-ELM	2.915	1.155	1.046	0.984	98.95%

The TCN-ELM hybrid model achieved a MAD of 1.552 MW, an RMSE of 2.915 MW, and a MAPE of 1.046%. Relative to LSTM, the best single deep model, the TCN-ELM, reduced errors by 68.2% in MAD, 46.5% in RMSE, and 54.6% in MAPE, while the R^2 increased from 0.943 to 0.984. While the ELM model performed well, its RMSE of 3.268 MW remained approximately 150% higher than that of the hybrid model, demonstrating the added benefit of the TCN front-end.

The TCN-ELM hybrid model demonstrated the highest accuracy (98.95%) and lowest RMSE (2.915 MW), proving to be the most effective model for forecasting wind power in Typical Daily Pattern 3. The TCN-ELM hybrid's ability to capture the low night-time power followed by a ramp to approximately 50 MW makes it an ideal choice for handling this type of wind power pattern. The model's performance underscores its utility in ultra-short-term wind power forecasting, real-time grid operations, and decision-making processes.

4.5. Deterministic Prediction Performance for Typical Daily Pattern 4

Four forecasting approaches (TCN, TCN-ELM, ELM, and LSTM) were analyzed for their predictive performance on Typical Daily Pattern 4, which was based on real wind farm measurements. DTW-based K-means clustering was adopted in the data preprocessing step to segment daily wind power trajectories, facilitating the discovery of distinct generation modes. Important features were selected using grey relational analysis, leading to enhanced input data quality through dimensionality optimization.

Cluster 4, as outlined in Table 1, exhibits a quasi-flat low-power trace ($\leq$25 MW) with sporadic gust spikes, indicative of weak wind conditions primarily influenced by frontal passages. These conditions, which are typical of weak wind backgrounds, were accurately captured by all the forecasting models, with the TCN-ELM hybrid model providing the highest accuracy in prediction.

Table 6 reports the deterministic errors obtained on the 15 min-ahead test set. Five classical criteria and the derived accuracy (%) are listed; all power metrics are normalized to megawatts (MW).

Table 6. Deterministic error metrics for 15 min-ahead wind-power forecasts on the test set (Cluster 4).

Prediction Model	RMSE/MW	MAD/MW	MAPE/%	R^2	Accuracy/%
TCN	7.794	6.169	4.157	0.751	95.83%
ELM	4.600	3.277	2.209	0.913	97.78%
LSTM	2.328	1.120	0.755	0.978	99.24%
TCN-ELM	1.606	0.752	0.507	0.989	99.49%

The TCN-ELM hybrid model achieved a MAD of 0.752 MW, an RMSE of 1.606 MW, and a MAPE of 0.507%, resulting in the highest accuracy of 99.49%. This represents a significant improvement over the standalone TCN model, which achieved a MAPE of 4.157% and an accuracy of 95.83%. Compared to the LSTM, the TCN-ELM reduced errors by 68.2% in MAD, 46.5% in RMSE, and 54.6% in MAPE, while increasing the R^2 value from 0.978 to 0.989.

The TCN-ELM hybrid model's performance underscores its strength in accurately predicting low-power periods typically observed during weak-wind conditions. It delivers the lowest RMSE and the highest accuracy, making it the most effective model for forecasting wind power in such regimes. Its superior ability to handle low-wind power dynamics, along with its robustness across varying intra-day conditions, makes it highly suitable for real-time operational forecasting and decision-making in wind power systems.

4.6. Enhancing TCN-ELM Forecasting Accuracy with DTW–K-Means Clustering

In Sections 4.1–4.5, the dataset was partitioned using DTW–K-means into four representative daily operating regimes. Within each regime, the TCN-ELM model was independently trained and evaluated, resulting in RMSE values between 1.6 MW and 5.7 MW (arithmetic mean $\approx$ 3.5 MW) and an average coefficient of determination of approximately 0.98. These results demonstrated that, once segmentation was applied, the convolutional encoder and the randomized ELM output layer were able to capture regime-specific temporal dynamics with high fidelity.

The application of DTW–K-means clustering was compared with that of traditional K-means clustering in ultra-short-term wind power forecasting. Traditional K-means clustering typically uses Euclidean distance to measure the similarity between data points, which does not account for the temporal variations and high volatility inherent in wind power data. In contrast, DTW–K-means, by employing DTW to calculate the similarity of time series, takes into consideration temporal variations and alignment and thus allows for a more accurate capture of operational patterns and temporal characteristics in wind power data.

Building on this regime-level performance, this section focuses on an experiment in which the DTW–K-means clustering stage was replaced by traditional K-means clustering, and a single TCN-ELM model was trained on the entire dataset. The errors resulting from this experiment were compared with those from the DTW–K-means clustered case to quantify the accuracy gains attributable to DTW–K-means segmentation in ultra-short-term wind power forecasting.

To visualize the quantitative impact of replacing DTW–K-means clustering with traditional K-means clustering, the principal deterministic error metrics obtained under the two experimental settings are presented in Table 7. This table contrasts the performance of the DTW–K-means clustered model with that of the traditional K-means clustered model, providing an at-a-glance assessment of the accuracy improvements achieved by DTW–K-means segmentation.

As seen in Table 7, the results indicate a marked improvement in the performance of the TCN-ELM model when DTW–K-means clustering is applied. The RMSE values for DTW–K-means clustering are consistently lower than those for traditional K-means clustering, with RMSE values of 3.700 MW (Cluster 1), 5.720 MW (Cluster 2), 2.915 MW (Cluster 3), and 1.606 MW (Cluster 4), compared to 10.229 MW (Cluster 1), 14.936 MW (Cluster 2), 16.570 MW (Cluster 3), and 19.604 MW (Cluster 4) for traditional K-means clustering.

Moreover, the accuracy of the TCN-ELM model using DTW–K-means clustering is significantly higher across all clusters, with accuracy ranging from 98.29% to 99.49%, compared to 88.01% to 93.91% for the model using traditional K-means clustering. These results demonstrate that DTW–K-means clustering not only enhances the accuracy of the model but also improves its robustness, especially in scenarios with high volatility and complex temporal patterns.

Table 7. Impact of clustering method on TCN-ELM model performance.

Cluster	Error Metric	DTW–K-Means Clustering	Traditional K-Means Clustering
1	RMSE/MW	3.700	10.229
	MAD/MW	2.532	9.016
	MAPE/%	1.706	6.076
	R^2	0.992	0.759
	Accuracy/%	98.29%	93.91%
2	RMSE/MW	5.720	14.936
	MAD/MW	3.884	11.695
	MAPE/%	2.617	7.881
	R^2	0.967	0.857
	Accuracy/%	97.38%	92.10%
3	RMSE/MW	2.915	16.570
	MAD/MW	1.155	14.198
	MAPE/%	1.046	9.567
	R^2	0.984	0.700
	Accuracy/%	98.95%	90.41%
4	RMSE/MW	1.606	19.604
	MAD/MW	0.752	17.739
	MAPE/%	0.507	11.953
	R^2	0.989	0.6412
	Accuracy/%	99.49%	88.01%

DTW–K-means clustering significantly improves model forecasting accuracy, particularly in operational regimes characterized by high volatility and complex temporal patterns. In contrast, traditional K-means clustering underperforms in these high-variability and complex patterns, as indicated by a significant increase in RMSE values and a decrease in accuracy. This verifies the advantage of DTW–K-means clustering in wind power forecasting. DTW–K-means enhances predictive accuracy by capturing the temporal alignment and dynamic changes in wind power data, particularly in high-volatility scenarios. In contrast, traditional K-means fails to account for temporal alignment, resulting in lower predictive accuracy, especially in high-variability and complex operational modes.

5. Conclusions

Addressing the critical challenge of ultra-short-term wind power forecasting, this research systematically compared the performance of TCN, ELM, LSTM, and the novel hybrid TCN-ELM model using real-world wind farm data. During the data preprocessing stage, an innovative clustering of daily wind power curves based on DTW-based K-means was performed to distinguish different generation patterns. Simultaneously, grey relational analysis was applied to select original features, effectively reducing dimensionality and enhancing input feature quality. Under the clustered setting, the hybrid TCN-ELM was found to deliver the lowest RMSE and capacity-normalized MAPE, while a coefficient of determination near 0.98 was maintained; when the clustering stage was omitted, deterministic errors increased markedly, thereby confirming the significant contribution of regime-aware modelling to predictive fidelity. Accordingly, it is indicated that coupling a deep convolutional encoder for temporal pattern extraction with a lightweight ELM output layer for rapid nonlinear mapping enables a balanced trade-off between accuracy and computational efficiency. Although the reported gains were observed on a specific dataset and may vary under different degrees of wind regime heterogeneity, the proposed method offers a practicable route toward more reliable real-time forecasting, thus providing technical support for grid dispatch, market participation, and wind farm operational scheduling. Future work could be extended by exploring adaptive clustering thresholds, integrating probabilistic postprocessing, and examining explainability tools to enhance model transparency and operator trust.

Author Contributions: Conceptualization, Y.Y. and Y.Z.; methodology, Y.Y. and Y.Z.; software, Y.Y. and Y.Z.; validation, Y.Y.; formal analysis, Y.Z.; investigation, Y.Z.; resources, Y.Y.; data curation, Y.Z.; writing—original draft preparation, Y.Y.; writing—review and editing, Y.Z.; visualization, Y.Y.; supervision, Y.Y.; project administration, Y.Y. and Y.Z.; funding acquisition, Y.Y. and Y.Z. All authors have read and agreed to the published version of the manuscript.

Funding: This work is supported by the Ningxia Natural Science Foundation Project, under Grant 2023AAC03836, and the Lianyungang City Key Research and Development Programme (Industrial Forward-Looking and Critical Core Technologies): Research on Ultra-Short-Term Probabilistic Forecasting of Photovoltaic Power Incorporating Micro-Meteorological Spatiotemporal Correlation (Project No. CG2315).

Data Availability Statement: The data that support the findings of this study are available on request from the corresponding author.

Conflicts of Interest: Author Yan Yan was employed by the State Grid Ningxia Electric Power Company. The remaining authors declare that the research was conducted in the absence of any commercial or financial relationships that could be construed as a potential conflict of interest.

References

1. Zhou, Y.; Sun, Y.; Wang, S.; Mahfoud, R.J.; Hou, D.; Wang, J. Very Short-Term Probabilistic Prediction for Regional Wind Power Generation Based on OPNPIs. *CSEE J. Power Energy Syst.* **2024**, *10*, 60–70. [CrossRef]
2. Yang, Y.; Lou, H.; Wu, J.; Zhang, S.; Gao, S. A Survey on Wind Power Forecasting with Machine Learning Approaches. *Neural Comput. Appl.* **2024**, *36*, 12753–12773. [CrossRef]
3. Yu, C.; Yan, G.; Yu, C.; Zhang, Y.; Mi, X. A Multi-Factor Driven Spatiotemporal Wind Power Prediction Model Based on Ensemble Deep Graph Attention Reinforcement Learning Networks. *Energy* **2023**, *263*, 126034.
4. Wang, S.; Zhang, W.; Sun, Y.; Trivedi, A.; Chung, C.Y.; Srinivasan, D. Wind Power Forecasting in the Presence of Data Scarcity: A Very Short-Term Conditional Probabilistic Modeling Framework. *Energy* **2024**, *291*, 130305. [CrossRef]
5. Zhou, Y.; Sun, Y.; Wang, S.; Mahfoud, R.J.; Alhelou, H.H.; Hatziargyriou, N.; Siano, P. Performance Improvement of Very Short-Term Prediction Intervals for Regional Wind Power Based on Composite Conditional Nonlinear Quantile Regression. *J. Mod. Power Syst. Clean Energy* **2022**, *10*, 60–70. [CrossRef]
6. Qin, R.; Chai, H.K.; Liu, K.; Yu, H.; Huang, J. A Hierarchical Multi-stage Fusion Deep Learning Framework for Short-Term Wind Power Prediction. *Renew. Energy* **2025**, *253*, 123551. [CrossRef]
7. Wang, Z.C.; Niu, J.C. Wind Power Output Prediction: A Comparative Study of Extreme Learning Machine. *Front. Energy Res.* **2023**, *11*, 1267275. [CrossRef]

8. Zhou, Y.; Wei, F.; Kuang, K.; Mahfoud, R.J. Research on A Deep Ensemble Learning Model for the Ultra-Short-Term Probabilistic Prediction of Wind Power. *Electronics* **2024**, *13*, 475. [CrossRef]
9. Xu, Y.; Wan, C.; Liu, H.; Zhao, C.; Song, Y. Probabilistic Forecasting-Based Reserve Determination Considering Multi-Temporal Uncertainty of Renewable Energy Generation. *IEEE Trans. Power Syst.* **2024**, *39*, 1019–1031. [CrossRef]
10. Wang, S.; Sun, Y.; Zhai, S.; Hou, D.; Wang, P.; Wu, X. Ultra-Short-Term Wind Power Forecasting Based on Deep Belief Network. In Proceedings of the 2019 Chinese Control Conference (CCC), Guangzhou, China, 27–30 July 2019; pp. 7479–7483.
11. Zhou, Y.; Sun, Y.; Wang, S.; Bai, L.; Hou, D.; Mahfoud, R.J.; Wang, P. Very Short-Term Probabilistic Prediction Method forWind Speed Based on ALASSO-Nonlinear Quantile Regression and Integrated Criterion. *CSEE J. Power Energy Syst.* **2023**, *9*, 2121–2129.
12. Kosanoglu, F. Wind Speed Forecasting with A Clustering-Based Deep Learning Model. *Appl. Sci.* **2022**, *12*, 13031. [CrossRef]
13. Sun, Y.; Zhou, Y.; Wang, S.; Mahfoud, R.J.; Alhelou, H.H.; Sideratos, G.; Hatziargyriou, N.; Siano, P. Nonparametric Probabilistic Prediction of Regional PV Outputs Based on Granule-Based Clustering and Direct Optimization Programming. *J. Mod. Power Syst. Clean Energy* **2023**, *11*, 1450–1461. [CrossRef]
14. Yang, M.; Guo, Y.; Fan, F. Ultra-Short-Term Prediction of Wind Farm Cluster Power Based on Embedded Graph Structure Learning with Spatiotemporal Information Gain. *IEEE Trans. Sustain. Energy* **2025**, *16*, 308–322. [CrossRef]
15. Wang, D.; Yang, M.; Zhang, W.; Ma, C.; Su, X. Short-Term Power Prediction Method of Wind Farm Cluster Based on Deep Spatiotemporal Correlation Mining. *Appl. Energy* **2025**, *380*, 125102. [CrossRef]
16. Yang, M.; Shen, X.; Huang, D.; Su, X. Fluctuation Classification and Feature Factor Extraction to Forecast Very Short-Term Photovoltaic Output Powers. *CSEE J. Power Energy Syst.* **2025**, *11*, 661–670.
17. Yang, M.; Jiang, Y.; Xu, C.; Wang, B.; Wang, Z.; Su, X. Day-Ahead Wind Farm Cluster Power Prediction Based on Trend Categorization and Spatial Information Integration Model. *Appl. Energy* **2025**, *388*, 125580. [CrossRef]
18. Yang, M.; Peng, T.; Zhang, W.; Su, X.; Han, C.; Fan, F. Abnormal Data Identification and Reconstruction Based on Wind Speed Characteristics. *CSEE J. Power Energy Syst.* **2025**, *11*, 612–622.
19. Niu, D.; Pu, D.; Dai, S. Ultra-Short-Term Wind-Power Forecasting Based on the Weighted Random Forest Optimized by the Niche Immune Lion Algorithm. *Energies* **2018**, *11*, 1098. [CrossRef]
20. Sun, Y.; Li, Z.; Yu, X.; Li, B.; Yang, M. Research on Ultra-Short-Term Wind Power Prediction Considering Source Relevance. *IEEE Access* **2020**, *8*, 147703–147710. [CrossRef]
21. Tan, B.; Ma, X.; Shi, Q.; Guo, M.; Zhao, H.; Shen, X. Ultra-Short-Term Wind Power Forecasting Based on Improved LSTM. In Proceedings of the 2021 6th International Conference on Power and Renewable Energy (ICPRE), 17–20 September 2021; pp. 1029–1033.
22. Wei, J.; Wu, X.; Yang, T.; Jiao, R. Ultra-Short-Term Forecasting of Wind Power Based on Multi-Task Learning and LSTM. *Int. J. Electr. Power Energy Syst.* **2023**, *149*, 109073. [CrossRef]
23. He, J.; Yu, C.; Li, Y.; Xiang, H. Ultra-Short Term Wind Prediction with Wavelet Transform, Deep Belief Network and Ensemble Learning. *Energy Convers. Manag.* **2020**, *205*, 112418. [CrossRef]
24. Xue, Y.; Yin, J.; Hou, X. Short-Term Wind Power Prediction Based on Multi-Feature Domain Learning. *Energies* **2024**, *17*, 3313. [CrossRef]
25. An, G.; Jiang, Z.; Cao, X.; Liang, Y.; Zhao, Y.; Li, Z. Short-Term Wind Power Prediction Based on Particle Swarm Optimization-Extreme Learning Machine Model Combined with AdaBoost Algorithm. *IEEE Access* **2021**, *9*, 94040–94052. [CrossRef]
26. Hao, J.; Zhu, C.; Guo, X. Wind Power Short-Term Forecasting Model Based on the Hierarchical Output Power and Poisson Re-sampling Random Forest Algorithm. *IEEE Access* **2021**, *9*, 6478–6487. [CrossRef]
27. Deng, W.; Dai, Z.; Liu, X.; Chen, R.; Wang, H.; Zhou, B. Short-Term Wind Power Prediction Based on Wind Speed Interval Division and TimeGAN for Gale Weather. In Proceedings of the 2023 International Conference on Power Energy Systems and Applications (ICoPESA), 24–26 February 2023; pp. 352–357.
28. Wang, L.; He, Y.; Li, L.; Liu, X.; Zhao, Y. A Novel Approach to Ultra-Short-Term Multi-Step Wind Power Predictions Based on Encoder–Decoder Architecture in Natural Language Processing. *J. Clean. Prod.* **2022**, *354*, 131723. [CrossRef]
29. Son, N.; Yang, S.; Na, J. Hybrid Forecasting Model for Short-Term Wind Power Prediction Using Modified Long Short-Term Memory. *Energies* **2019**, *12*, 3901. [CrossRef]
30. Meng, Y.; Chang, C.; Huo, J.; Zhang, Y.; Mohammed Al-Neshmi, H.M.; Xu, J.; Xie, T. Research on Ultra-Short-Term Prediction Model of Wind Power Based on Attention Mechanism and CNN-BiGRU Combined. *Front. Energy Res.* **2022**, *10*, 920835. [CrossRef]
31. Ju, Y.; Sun, G.; Chen, Q.; Zhang, M.; Zhu, H.; Rehman, M.U. A Model Combining Convolutional Neural Network and LightGBM Algorithm for Ultra-Short-Term Wind Power Forecasting. *IEEE Access.* **2019**, *7*, 28309–28318. [CrossRef]
32. Yin, H.; Ou, Z.; Huang, S.; Meng, A. A Cascaded Deep Learning Wind Power Prediction Approach Based on A Two-Layer of Mode Decomposition. *Energy* **2019**, *189*, 116316. [CrossRef]

33. Wang, K.; Qi, X.; Liu, H.; Song, J. Deep Belief Network Based K-Means Cluster Approach for Short-Term Wind Power Forecasting. *Energy* **2018**, *165*, 840–852. [CrossRef]
34. Zhang, S.; Zhu, C.; Guo, X. Wind-Speed Multi-Step Forecasting Based on Variational Mode Decomposition, Temporal Convolutional Network, and Transformer Model. *Energies* **2024**, *17*, 1996. [CrossRef]
35. Li, Q.; Ren, X.; Zhang, F.; Gao, L.; Hao, B. A Novel Ultra-Short-Term Wind Power Forecasting Method Based on TCN and Informer Models. *Comput. Electr. Eng.* **2024**, *120*, 109632. [CrossRef]
36. Wu, W.; Peng, M. A Data Mining Approach Combining K-Means Clustering with Bagging Neural Network for Short-Term Wind Power Forecasting. *IEEE Internet Things J.* **2017**, *4*, 979–986. [CrossRef]
37. Liu, S.; Xu, T.; Du, X.; Zhang, Y.; Wu, J. A Hybrid Deep Learning Model Based on Parallel Architecture TCN-LSTM with Savitzky-Golay Filter for Wind Power Prediction. *Energy Convers. Manag.* **2024**, *302*, 118122. [CrossRef]
38. Medina, S.V.; Ajenjo, U.P. Performance Improvement of Artificial Neural Network Model in Short-Term Forecasting of Wind Farm Power Output. *J. Mod. Power Syst. Clean Energy* **2020**, *8*, 484–490. [CrossRef]
39. Yildiz, C.; Acikgoz, H.; Korkmaz, D.; Budak, U. An Improved Residual-Based Convolutional Neural Network for Very Short-Term Wind Power Forecasting. *Energy Convers. Manag.* **2021**, *228*, 113731. [CrossRef]
40. Zhang, S.; Chen, Y.; Xiao, J.; Zhang, W.; Feng, R. Hybrid Wind Speed Forecasting Model Based on Multivariate Data Secondary Decomposition Approach and Deep Learning Algorithm with Attention Mechanism. *Renew. Energy* **2021**, *174*, 688–704. [CrossRef]
41. Wang, L.; Tao, R.; Hu, H.; Zeng, Y.R. Effective Wind Power Prediction Using Novel Deep Learning Network: Stacked Independently Recurrent Autoencoder. *Renew. Energy* **2021**, *164*, 642–655. [CrossRef]
42. Yang, M.; Ju, C.; Huang, Y.; Guo, Y.; Jia, M. Short-Term Power Forecasting of Wind Farm Cluster Based on Global Information Adaptive Perceptual Graph Convolution Network. *IEEE Trans. Sustain. Energy* **2024**, *15*, 2063–2076. [CrossRef]

energies

MDPI

Article

Differentiated Pricing-Mechanism Design for Renewable Energy with Analytical Uncertainty Representation

Xianzhuo Liu [1], Xue Yuan [2,*], Qi An [2] and Jiale Liu [1]

[1] China Southern Power Grid Dispatching and Control Center, Guangzhou 510663, China; liu-xz14@tsinghua.org.cn (X.L.); 202111082844@mail.scut.edu.cn (J.L.)
[2] State Key Laboratory of Alternate Electrical Power System with Renewable Energy Sources, North China Electric Power University, Beijing 102206, China; anqi-ncepu@foxmail.com
* Correspondence: xueyuanncepu@163.com

Abstract

With the integration of high-penetration renewable energy, existing uniform marginal pricing mechanisms face critical challenges, including difficulty in recovering flexibility resource capacity costs and free-riding phenomena caused by renewable energy's variability. To address these issues, this paper proposes a differentiated pricing mechanism for renewable energy based on analytical uncertainty representation to avoid marginal price distortion and promote the rational allocation of ancillary service costs. Firstly, a joint clearing model for energy and reserve ancillary service is developed, incorporating a distributional robust chance constraint based on moment information to model the uncertainty of renewable energy. Then, the composition structure of the nodal marginal price for ancillary service demand is redefined, offering clearer and more explicit price signals compared with traditional uniform marginal pricing. After that, quantification of the impact of energy storage on renewable energy forecast errors and ancillary service pricing is conducted, with a systematic analysis of its role in reducing ancillary service costs and optimizing generation revenue. Simulation results on the modified IEEE 30-bus system demonstrate significant advantages over traditional uniform pricing: the proposed mechanism ensures fair cost allocation, effectively mitigates free-riding problems, and provides clear economic signals. With energy storage units regulating renewable power output, it could lead to a 12.9% reduction in ancillary service costs while increasing total generation revenue by 6.73%.

Keywords: ancillary service; differentiated pricing; high proportion of renewable energy; market clearing; uncertainty distribution

1. Introduction

1.1. Background

With the global transition of the energy structure and the continuous deepening of electricity market reforms, new energy forms, such as photovoltaic, wind power, and micro-grids, have gradually become important market participants, playing an increasingly significant role in electricity trading [1,2]. Since the government's cancelation of subsidy policies for renewable energy sources in 2021, the era of grid parity for renewable energy has begun, and electricity prices are now set by market forces [3]. Under the current marginal cost pricing mechanism adopted in the electricity spot market, promoting a high proportion of renewable energy participation in market transactions has posed new challenges for electricity market construction, primarily reflected in the following two aspects.

On one hand, price signals are distorted. Large-scale renewable integration suppresses average electricity prices and amplifies volatility, posing structural challenges for both renewable energy and thermal power [4]. With near-zero marginal costs, renewable energy acts as price-makers during periods of abundant output, driving prices sharply downward—even to zero or negative levels [5,6]. Conversely, during renewable shortages, thermal plants become marginal units, triggering short-term price spikes and exacerbating market instability. While renewable energy dominates bidding due to low marginal costs, prolonged price depression erodes their revenue streams, thereby undermining investment returns. Meanwhile, thermal power transitions from baseload to flexible operation, facing reduced utilization rates, challenges in recovering fixed costs, and potential premature retirement due to insufficient revenue, further degrading system flexibility [7,8].

On the other hand, current marginal pricing mechanisms fail to account for differences in renewable uncertainty. Renewable energy with high uncertainty requires greater flexibility support but receive the same pricing as low-uncertainty counterparts, which fails to internalize their system-level externalities. Current mechanisms do not allocate ancillary service costs proportionally based on volatility differences, resulting in cost misallocation and free-riding problems [9]. These issues weaken incentives for renewable plants to improve forecasting accuracy and hinder the sustainable development of electricity markets.

Therefore, it is imperative to introduce uncertainty-sensitive market mechanisms, overhaul the uniform marginal-cost pricing model, and deliver more precise price signals. Such reforms will ensure market prices reflect the true value of generation, safeguard the economic viability of flexible resources like thermal power, and enhance both market efficiency and long-term sustainability.

1.2. Literature Review

Currently, domestic and international electricity markets are actively advancing reforms in ancillary service markets. For instance, Gansu Province determines frequency regulation prices through competitive bidding, allowing renewable energy to declare ancillary service demands based on forecast deviations and procure required capacity via market mechanisms [10]. Ref. [11] proposes a market-clearing model where renewable energy actively declares reserve demands, bearing reserve capacity costs based on declared quantities and the cleared marginal price. However, while such mechanisms allocate costs according to renewable energy' ancillary service demands, the uniform marginal clearing price overlooks price distortions caused by stations with large forecast deviations. This forces low-volatility stations to subsidize high-volatility ones, worsening free-riding and undermining market fairness. Ref. [12] introduces a frequency regulation cost-allocation method based on volatility risk, allocating costs according to each participant's maximum forecast deviation as a share of total deviations. Although simple and intuitive, this approach fails to distinguish deviation sources, resulting in imprecise cost allocation, ineffective price signals, and insufficient incentives for renewable energy to improve forecasting accuracy. To address the risks of energy imbalance in power systems, Germany and the UK distribute ancillary service costs through balancing mechanisms and imbalance settlements [13]. In the U.S. PJM market, flexibility resources receive differentiated compensation based on response speed, precision, and sustainability, quantified through performance coefficients, with a two-part settlement (capacity payment + performance payment) to incentivize operational optimization [14]. Yet, this mechanism still charges renewable energy a uniform unit price for ancillary services, disregarding their varying impacts on system flexibility due to uncertainty.

With the increasing penetration of renewable energy, researchers have begun to explore pricing mechanisms that address renewable energy uncertainty. The two dominant approaches are robust optimization and chance constraints. Ref. [15] derives the price components of renewable energy uncertainty based on its extreme output values but does not explicitly consider its probabilistic distribution characteristics, which may lead to distorted price signals. Ref. [16] proposes a robust economic dispatch model and introduces Uncertainty Marginal Pricing (UMP) to quantify the cost of renewable energy uncertainty. Under this mechanism, renewable energy units with a higher-output volatility are required to pay higher fees, with the collected funds used to compensate market participants providing reserve capacity. However, robust optimization methods tend to be overly conservative, which may restrict market flexibility and hinder renewable energy integration. To address the conservatism of robust optimization, Ref. [17] introduces a chance-constrained frequency-regulation marginal pricing model, using probabilistic constraints to improve the economic efficiency of frequency regulation markets. Ref. [18] adopts a chance-constrained stochastic optimization approach to develop an electricity market pricing mechanism that explicitly considers renewable energy uncertainty based on multiple statistical moments. However, this method relies heavily on extensive historical data to characterize the probability distribution of renewable generation, which is often challenging to obtain in practice, limiting its feasibility. Building on these approaches, Ref. [19] proposes a market-clearing model based on distributional robust chance constraints, which simultaneously optimizes energy, reserves, and uncertainty resources. Unlike conventional stochastic models, this approach eliminates the need for precise probabilistic distribution assumptions, thereby enhancing adaptability and robustness. Ref. [20] and Ref. [21] further develop a Wasserstein distributional robust optimization approach to enhance uncertainty pricing. However, the high computational complexity of this method may constrain its scalability in large-scale market applications. A comparison of robust optimization, chance-constrained methods, and distributional robust chance constraints is shown in Table 1.

Table 1. Comparison of three methods.

Approach	Key Features	Assumptions	Limitations
Robust Optimization	Price uncertainty via worst-case bounds.	Constraints must be linear or convex with respect to uncertain parameters.	Ignores probabilistic behavior
Chance-Constrained Methods	Uses probabilistic constraints for efficiency.	Requires extensive historical data.	High computational cost
Distributional Robust Chance Constraints	Uncertainty of random variables and probability distributions.	Sample independence for confidence level calculation	Conservativeness depends on fuzzy sets

1.3. Knowledge Gap

Overall, existing pricing methods fail to directly price ancillary services from the demand side, focusing primarily on the uncertainty components of renewable energy while lacking a clear linkage between responsibility and cost. This limits the effective participation of flexibility resources in the market and may lead to market distortions. Moreover, the impact of energy storage allocation on the pricing mechanism for renewable energy uncertainty remains underexplored. Based on the above analysis, this paper proposes a differentiated pricing mechanism for renewable energy based on analytical uncertainty representation, which could avoid marginal price distortion and promote the rational allocation of ancillary service costs without the limitations of existing methods.

1.4. Contribution

Key contributions of this paper are as follows:

(1) A joint clearing model for energy and reserve ancillary services is developed, incorporating a distributionally robust chance constraint based on moment information to model the uncertainty of renewable energy. The composition structure of the nodal marginal price for ancillary service demand is redefined, offering clearer and more explicit price signals compared with traditional uniform marginal pricing. This approach ensures a fair allocation of ancillary service costs, effectively mitigating the free-riding problem and reducing excessive ancillary service capacity allocation.

(2) Quantification of the impact of energy storage on renewable energy forecast errors and ancillary service pricing is conducted, with a systematic analysis of its role in reducing ancillary service costs and optimizing generation revenue. The results demonstrate that the proposed mechanism incentivizes renewable energy plants to deploy energy storage to regulate output uncertainty, improve predictability and market competitiveness, and facilitate the stable and efficient integration of high-penetration renewable energy into the power grid.

The rest of this paper is organized as follows: In Section 2, the methodology of this paper is introduced. In Section 3, the basic principle of differentiated pricing is illustrated. In Section 4, a joint clearing model for energy and reserve ancillary services is developed, incorporating distributional robust chance constraints based on moment information to handle renewable energy uncertainty. In Section 5, the nodal marginal price structure is redefined by decomposing ancillary service prices into energy and congestion components for differentiated pricing. In addition, bi-level optimization quantitatively analyzes energy storage's impact on forecasting errors and pricing. In Section 6, using the modified IEEE 30-bus system, a comparative analysis between traditional uniform pricing and uncertainty-aware differentiated pricing validates significant advantages in cost allocation fairness and market incentives. Section 7 concludes this paper.

2. Methodology

The methodology of this paper is as follows: This paper aims to study a differentiated pricing mechanism for ancillary service demand based on the uncertainties of various renewable energy sources, reasonably quantifying the ancillary service costs caused by the integration of different renewable energy types. This approach promotes the rational optimization of resources and enhances market competitiveness. The research framework is illustrated in Figure 1.

To accurately reflect the supply–demand relationship and the true costs of power generation entities, this paper constructs a joint clearing model for the electricity and ancillary service markets, considering the impact of renewable energy uncertainty on system reserve capacity requirements and flow margins. Based on the statistical characteristics of renewable energy output errors, reserve transmission line opportunity constraints and generation reserve opportunity constraints are established. A joint clearing of electricity and ancillary services is conducted, and the ancillary service demand price for renewable energy is derived using Lagrangian duality. This approach reflects the differentiated stochastic and spatial characteristics of renewable energy, conveying reasonable price signals and guiding renewable energy sources to bear the ancillary service costs they cause. Renewable energy entities declare their demand based on their output characteristics, which helps achieve precise matching of supply and demand and reduces system operation costs.

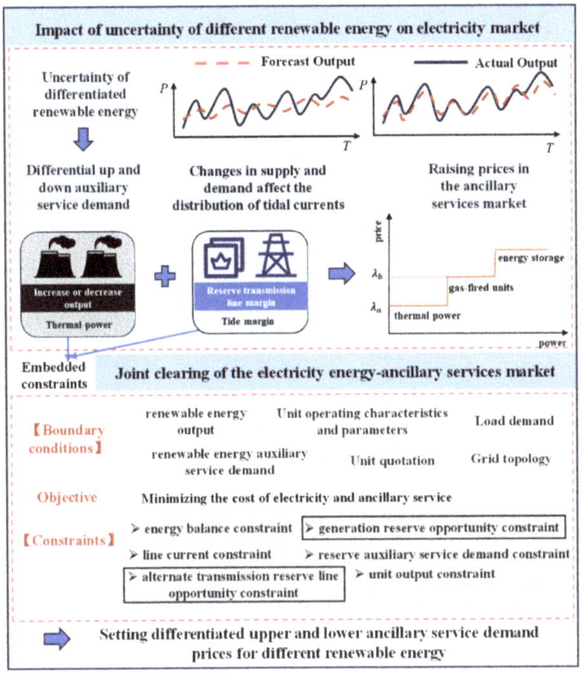

Figure 1. Demand-differentiated pricing-framework diagram for ancillary services based on renewable energy uncertainty.

3. Basic Principle

Renewable energy sources exhibit varying degrees of uncertainty, measured by forecast error statistics. The forecast error is defined as follows:

$$\Delta P_i^{\mathrm{W}} = P_{\mathrm{act},i}^{\mathrm{W}} - P_{\mathrm{plan},i}^{\mathrm{W}} \tag{1}$$

where ΔP_i^{W} represents the forecast error of renewable energy generation; $P_{\mathrm{act},i}^{\mathrm{W}}$ represents the actual generation output of renewable energy; $P_{\mathrm{plan},i}^{\mathrm{W}}$ represents the forecasted generation of renewable energy.

As shown in Figure 2, negative forecast errors (actual output < forecasted) create upward reserve demand, requiring flexible resources to increase output and fill the power gap. Positive forecast errors (actual output > forecasted) generate downward reserve demand, necessitating output reduction to prevent excess supply.

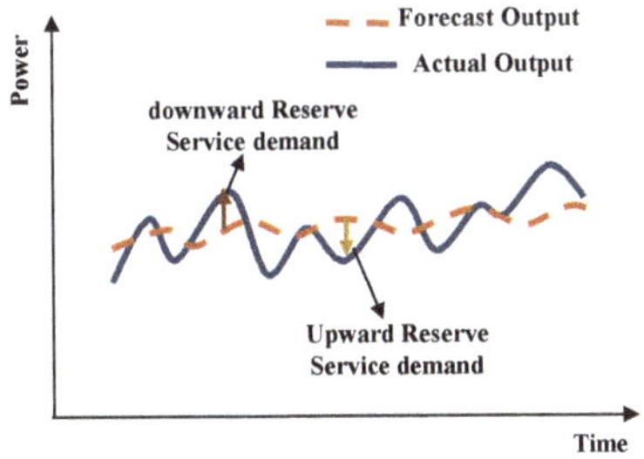

Figure 2. Renewable energy forecast and actual output curves.

The prevailing uniform marginal pricing scheme leads to market distortions, as ancillary service demand increments at any renewable energy station result in cross-subsidization from other participants. This violates incentive compatibility principles and fails to provide effective signals for uncertainty reduction.

The traditional uniform marginal price model fails to reflect the differences in the random fluctuations of various renewable energy sources. This highlights the urgent need for an ancillary service demand pricing mechanism that accounts for renewable energy uncertainty. This mechanism would provide reasonable ancillary service demand prices to represent the security costs associated with renewable energy sources.

The relationship between forecast errors and ancillary costs is analyzed below. In the ancillary service market-clearing model, forecast errors are used to quantify renewable uncertainty and corresponding ancillary costs. These costs must be rationally allocated to incentivize renewable energy to mitigate uncertainty.

4. Joint Clearing Model for Energy and Ancillary Service

This section focuses on studying the uncertainty of renewable energy sources. It is assumed that the load has no uncertainty, and network losses are neglected. A DC power flow model is used to describe the system's power balance and network security constraints.

4.1. Objective Function

The objective function is to minimize the total system operating cost, with the generation output and up/down spinning reserves as decision variables. The expression is as follows:

$$\min C = C_1 + C_2 \tag{2}$$

$$C_1 = \sum_{i \in I_G} \left(a_i \left(P_i^{G} \right)^2 + b_i P_i^{G} + c_i \right) \tag{3}$$

$$C_2 = \sum_{i \in I_G} \left(m_i^{up} r_i^{Gup} + m_i^{dn} r_i^{Gdn} \right) \tag{4}$$

where C_1 represents the fuel cost of thermal power units; C_2 represents the cost of procuring reserve ancillary services; I_G represents the set of thermal power units connected to node i; a_i, b_i, and c_i are the fuel cost coefficients; m_i^{up} and m_i^{dn} indicate the declared prices for providing upward and downward reserves by the thermal power unit, respectively; r_i^{Gup} and r_i^{Gdn} represent the cleared upward and downward reserve capacities of the thermal power unit i, respectively.

4.2. Constraints

(1) Power balance constraint:

$$\sum_{i \in I_G} P_i^{G} + \sum_{i \in I_W} P_i^{W} = \sum_{i \in I_D} D_i : \lambda_e \tag{5}$$

where I_W and I_D represent the sets of wind power units and loads connected to node, respectively; P_i^{W} is the day-ahead scheduled generation power of wind power unit i; D_i is the load power at node i; λ_e is the Lagrange multiplier corresponding to Equation (5).

(2) Line power flow constraints:

$$f_l = \sum_{i \in I} PTDF_{l-i} (P_i^{G} + P_i^{W} - D_i) \tag{6}$$

$$-F_l \leq f_l \leq F_l : \mu_l^{\min}, \mu_l^{\max} \tag{7}$$

where f_l represents the power flow on line l; I represents system nodes; $PTDF_{l-i}$ is the power transfer distribution factor of branch l with respect to node i; F_l represents the maximum transmission capacity of line l; $\mu_l^{\max}$ and $\mu_l^{\min}$ are the Lagrange multipliers corresponding to Equation (7).

(3) Reserve transmission line reservation opportunity constraints:

$$\Pr(f_l - \sum_{i \in I_W} PTDF_{l-i} r_i^{\mathrm{Wup}} + \sum_{i \in I_G} PTDF_{l-i} \Delta P_i^{\mathrm{up}} \leq F_l) \geq 1 - \varepsilon : \overline{\mu}_l^{\mathrm{ru}} \tag{8}$$

$$\Pr(-F_l \leq f_l - \sum_{i \in I_W} PTDF_{l-i} r_i^{\mathrm{Wup}} + \sum_{i \in I_G} PTDF_{l-i} \Delta P_i^{\mathrm{up}}) \geq 1 - \varepsilon : \underline{\mu}_l^{\mathrm{ru}} \tag{9}$$

$$\Pr(f_l + \sum_{i \in I_W} PTDF_{l-i} r_i^{\mathrm{Wdn}} - \sum_{i \in I_G} PTDF_{l-i} \Delta P_i^{\mathrm{dn}} \leq F_l) \geq 1 - \varepsilon : \overline{\mu}_l^{\mathrm{rd}} \tag{10}$$

$$\Pr(-F_l \leq f_l + \sum_{i \in I_W} PTDF_{l-i} r_i^{\mathrm{Wdn}} - \sum_{i \in I_G} PTDF_{l-i} \Delta P_i^{\mathrm{dn}}) \geq 1 - \varepsilon : \underline{\mu}_l^{\mathrm{rd}} \tag{11}$$

where Pr() represents the probability of satisfying the conditions within the parentheses; r_i^{Wup} and r_i^{Wdn} denote the upward and downward reserve requirements declared by the wind power unit, respectively; ΔP_i^{up} and ΔP_i^{dn} indicate the upward and downward generation adjustments for the thermal power unit i, respectively; ε represents the acceptable risk level of the chance constraint; $\overline{\mu}_l^{\mathrm{ru}}$ and $\underline{\mu}_l^{\mathrm{ru}}$ are the Lagrange multipliers corresponding to Equations (8) and (9); $\overline{\mu}_l^{\mathrm{rd}}$ and $\underline{\mu}_l^{\mathrm{rd}}$ are the Lagrange multipliers corresponding to Equations (10) and (11). Equations (8) through (11) consider the impact of ancillary services on line power flow, with capacity margins required on the line to address the uncertainty of renewable energy.

(4) Reserve ancillary service demand constraints:

$$\sum_{i \in I_G} r_i^{\mathrm{Gup}} \geq \sum_{i \in I_W} r_i^{\mathrm{Wup}} \tag{12}$$

$$\sum_{i \in I_G} r_i^{\mathrm{Gdn}} \geq \sum_{i \in I_W} r_i^{\mathrm{Wdn}} \tag{13}$$

where Equations (12) and (13) represent the upward and downward reserve requirement constraints declared by renewable energy sources, respectively. The total upward reserve provided by thermal power units must be greater than or equal to the total upward reserve capacity required by renewable units, and the total downward reserve provided must be greater than or equal to the total downward reserve capacity required by renewable units.

(5) Deviation balance adjustment constraints:

$$\Delta P_i^{\mathrm{up}} = \beta_i \sum_{i \in I_W} r_i^{\mathrm{Wup}} \tag{14}$$

$$\sum_{i \in I_G} \beta_i = 1 \tag{15}$$

$$\Delta P_i^{\mathrm{dn}} = \gamma_i \sum_{i \in I_W} r_i^{\mathrm{Wdn}} \tag{16}$$

$$\sum_{i \in I_G} \gamma_i = 1 \tag{17}$$

where β_i and γ_i represent the balancing coefficients for the upward and downward reserve requirements, respectively, defining the proportion of response allocated to each thermal unit in the regulation reserve. Higher balancing coefficients indicate that the corresponding units are required to undertake a larger share of the regulatory power. Constraints (15) and (17) indicate that the equilibrium coefficients sum to 1, balancing the power deviation caused by the renewable energy stochastic volatility.

(6) Generation reserve opportunity constraints:

$$\Pr\left(r_i^{\mathrm{Gup}} \geq \Delta P_i^{\mathrm{up}}\right) \geq 1-\varepsilon : \eta_{\mathrm{R},i}^{\mathrm{Gup}} \tag{18}$$

$$\Pr\left(r_i^{\mathrm{Gdn}} \geq \Delta P_i^{\mathrm{dn}}\right) \geq 1-\varepsilon : \eta_{\mathrm{R},i}^{\mathrm{Gdn}} \tag{19}$$

where Equations (18) and (19) represent the chance constraints for the upward and downward spinning reserves provided by thermal power units, respectively; $\eta_{\mathrm{R},i}^{\mathrm{Gup}}$ and $\eta_{\mathrm{R},i}^{\mathrm{Gdn}}$ are the Lagrange multipliers corresponding to these constraints.

(7) Thermal power unit output constraints:

$$P_i^{\mathrm{G}} + r_i^{\mathrm{Gup}} \leq P_{\mathrm{max},i} \tag{20}$$

$$-P_i^{\mathrm{G}} + r_i^{\mathrm{Gdn}} \leq -P_{\mathrm{min},i} \tag{21}$$

where $P_{\mathrm{max},i}$ and $P_{\mathrm{min},i}$ represent the upper and lower output limits of the thermal power unit i, respectively.

(8) Renewable energy unit output constraint:

$$0 \leq P_i^{\mathrm{W}} \leq P_i^{\mathrm{Wfcst}} \tag{22}$$

where P_i^{Wfcst} is the forecast generation power of wind power unit i.

4.3. Model Solution

Equations (8)–(11), (18), and (19) represent chance constraints, which are non-convex and challenging to solve directly. Typically, they need to be transformed into a deterministic model for a solution. This paper adopts a data-driven distributional robust optimization approach to transform and solve these constraints, which not only fully leverages the distribution characteristics of renewable energy output errors but also ensures the robustness of the optimization results [19].

First, assume a fuzzy set Φ_{mo} for the probability distribution of wind power output errors.

$$\Phi_{mo} = \left\{ P(\xi) \left| \begin{array}{l} \mathrm{E}_P(\xi) = \mu \\ \mathrm{E}_P\left[(\xi-\mu)(\xi-\mu)^T\right] = \Sigma \end{array} \right. \right\} \tag{23}$$

where ξ represents a random variable with an uncertain distribution, specifically the forecast error of renewable energy; $P(\cdot)$ denotes the probability density function of the random variable ξ; $E_P(\cdot)$ represents the expectation of the random variable ξ; μ and $\sum$ denote the mean vector and covariance matrix of the renewable energy forecast error, respectively.

The distributional robust chance constraint can be expressed as follows:

$$\inf_{P(\xi)\in\Phi_{mo}} \Pr(g(x,\xi) \leq 0) \geq 1-\varepsilon \tag{24}$$

where x is a decision variable.

For a linear constraint, the distributionally robust chance constraint can be written as follows:

$$g(x,\xi) = a(x)^T\xi + b(x) \leq 0 \tag{25}$$

$$\inf_{P(\xi)\in\Phi_{mo}} \Pr(a(x)^T\xi + b(x) \leq 0) \geq 1 - \varepsilon \tag{26}$$

Equation (26) can thus be transformed into a convex second-order cone constraint.

$$a(x)^T\mu + b(x) + \kappa_\varepsilon\sqrt{a(x)^T \sum a(x)} \leq 0 \tag{27}$$

where κ_ε is a constant that represents the transfer factor of the forecast error's standard deviation. The larger its value, the more conservative the economic dispatch scheme becomes. Under distributional robustness, the transfer factor is calculated as follows:

$$\kappa_\varepsilon = \sqrt{(1-\varepsilon)/\varepsilon} \tag{28}$$

For detailed analysis and transformation of the chance constraints on transmission reserve lines and generation reserves, see Appendix A.1.

4.4. Energy Storage Configuration

Energy storage can effectively mitigate the variability of renewable energy, maintain the generation-load balance in the new power system, and ensure system security and stability. The energy storage is deployed by renewable energy to optimize its behaviors of charging and discharging, thereby reducing the renewable energy forecast error. With the decrease in forecast error, the reserve provided by thermal power units is reduced, thereby decreasing the corresponding ancillary service costs and prices, which could effectively improve system operation flexibility and reduce system operation cost. By configuring energy storage devices at renewable plants, power output can be smoothed, reducing random fluctuations and thus decreasing the required ancillary service costs. To analyze changes in forecast error and ancillary service demand pricing at renewable energy stations after configuring energy storage, this paper constructs a bi-level model. The upper-level model is an optimization model for forecasting deviations in renewable energy with storage configuration, aiming to minimize forecast error after integrating storage. The lower-level model is the previously introduced market-clearing model, targeting the minimization of the system's total cost. This section mainly introduces the upper-level model, while the content of the lower-level model remains consistent with the previous discussion. The objective function of the upper-level model is as follows:

$$\min \varepsilon_s \tag{29}$$

$$\varepsilon_s = \left| P_{\text{forecast}} - P_{\text{actual}} + \left(P_k^{\text{dis}} - P_k^{\text{cha}}\right)\right| \tag{30}$$

where ε_s is the slack variable for the forecast error; P_{forecast} is the forecasted output of renewable energy; P_{actual} is the actual output of renewable energy; P_k^{cha} and P_k^{dis} represent the charging and discharging power of the energy storage, respectively. To reduce forecast error, when the forecasted output of renewable energy exceeds the actual output, the energy storage discharges; when the forecasted output is lower than the actual output, the energy storage charges.

The constraints of the upper-level model include the upper and lower limits for energy storage charging and discharging, as well as mutually exclusive constraints for charging and discharging behaviors.

$$0 \le U_k^{\text{cha}} + U_k^{\text{dis}} \le 1 \tag{31}$$

$$0 \le P_k^{\text{cha}} \le U_k^{\text{cha}} P_k^{\text{inv}} \tag{32}$$

$$0 \le P_k^{\text{dis}} \le U_k^{\text{dis}} P_k^{\text{inv}} \tag{33}$$

where U_k^{cha} and U_k^{dis} are indicative of energy storage charging and discharging states; U_k^{dis} P_k^{inv} is the planned rated power of the energy storage P_k^{inv} k.

The adjusted outputs of renewable energy, along with the mean and standard deviation of the forecast error, are passed to the lower-level model for market clearing.

$$P_s = P_{\text{forecast}} + P_k^{\text{dis}} \cdot \eta - \frac{P_k^{\text{cha}}}{\eta} \tag{34}$$

$$\mu_s = \frac{1}{N} \sum_{t=1}^{N} \varepsilon_s \tag{35}$$

$$\sigma_s = \sqrt{\frac{1}{N} \sum_{t=1}^{N} (\varepsilon_s - \mu_s)^2} \tag{36}$$

where P_s represents the overall power output of the renewable energy station after energy storage configuration; μ_s and σ_s are the mean and standard deviation of the renewable energy forecast error after integrating energy storage.

The bi-level model employs an iterative substitution approach to achieve equilibrium between integrated markets and energy storage systems. The upper-level model optimizes energy storage configuration by minimizing forecast errors and coordinating renewable energy output and storage behaviors, then passes these decisions to the lower-level model. The lower-level market-clearing model computes reserve requirements and electricity prices, feeding price signals back to the upper level. Through this iterative interaction, the equilibrium between joint market and energy storage is realized.

5. Clearing Price and Settlement Mechanism

5.1. Energy Market

In the energy market, the locational marginal price at a node is the increase in the total system cost caused by adding a unit load at that node, and it can be expressed as follows:

$$\begin{aligned} \lambda_i = \frac{\partial L}{\partial D_i} = \; & \lambda_e + \sum_{i \in I} PTDF_{l-i}(\mu_l^{\min} - \mu_l^{\max}) \\ & + \sum_{i \in I} PTDF_{l-i}\left(\underline{\mu}_l^{\text{ru}} - \overline{\mu}_l^{\text{ru}} + \underline{\mu}_l^{\text{rd}} - \overline{\mu}_l^{\text{rd}}\right) \end{aligned} \tag{37}$$

where λ_i represents the locational marginal price at node i. The LMP consists of three components: the first term, λ_e, is the energy component; the second term refers the conventional line congestion component, $\mu_l^{\max}$ and $\mu_l^{\min}$, which are the Lagrange multipliers for the upper and lower limits of line power flow transmission, respectively; and the third term is the congestion component for reserved transmission lines, $\overline{\mu}_l^{\text{ru}}$ and $\underline{\mu}_l^{\text{ru}}$, which are the Lagrange multipliers for the upper and lower constraints of line power flow when considering upward reserve requirements, and $\overline{\mu}_l^{\text{rd}}$ and $\underline{\mu}_l^{\text{rd}}$ are the Lagrange multipliers for the upper and lower constraints of line power flow when considering downward reserve requirements.

The settlement for each market participant in the energy market is as follows:

(1) Revenue for thermal power units:

$$F_{\text{Gen}} = \sum_{m \in I_{\text{G}}} \lambda_m P_m^{\text{G}} \tag{38}$$

(2) Revenue for wind power units:

$$F_{\text{Wind}} = \sum_{w \in I_{\text{W}}} \lambda_w P_w^{\text{W}} \tag{39}$$

(3) Payment by loads:

$$F_{\text{Load}} = \sum_{d \in I_{\text{D}}} \lambda_d D_d \tag{40}$$

where λ_m, λ_w, and λ_d represent the locational marginal prices at the nodes where thermal power units m, wind power units w, and loads d are located, respectively.

5.2. Ancillary Service Market

Theorem: The reserve demand price for wind power consists of two components: the reserve demand energy component and the reserve demand congestion component, which represents the scarcity value of the reserve transmission capacity and reflects the impact of the reserve transmission capacity limit on the reserve price. The reserve demand congestion component is directly related to congestion surplus, which refers to the economic revenue due to the price differential in the reserve market caused by reserve transmission congestion. The proof can be found in Appendix A.2.

Taking the above reserve demand price as an example, the price components are as follows:

$$\lambda_{r_i^{\text{Wup}}} = \lambda_{\text{rup},i}^{\text{EN}} + \lambda_{\text{rup},i}^{\text{CO}} \tag{41}$$

$$\lambda_{\text{rup},i}^{\text{EN}} = \sum_{i \in I_{\text{G}}} \kappa_\varepsilon \eta_{\text{R},i}^{\text{Gup}} \frac{\beta_i{}^2 \xi_{w,i}{}^2 r_i^{\text{Wup}} + \beta_i \xi_{w,i} \sum_{j \in I_{\text{W}}} \rho_{w,i,j} \beta_{w,j} \xi_j r_i^{\text{Wup}}}{\sqrt{K_{\text{RG,up}}{}^T \Sigma_{\text{W}} K_{\text{RG,up}}}} \tag{42}$$

$$\lambda_{\text{rup},i}^{\text{CO}} = \sum_{l \in I} \left[\frac{\kappa_\varepsilon (\overline{\mu}_l^{\text{ru}} + \underline{\mu}_l^{\text{ru}}) \times K_{\text{PF,up},li}{}^2 \xi_{w,i}{}^2 r_i^{\text{Wup}} + K_{\text{PF,up},li} \xi_{w,i} \sum_{j \in I_{\text{W}}} \rho_{w,i,j} K_{\text{PF,up},lj} \xi_{w,j} r_i^{\text{Wup}}}{\sqrt{K_{\text{PF,up}}{}^T \Sigma_{\text{W}} K_{\text{PF,up}}}} \right] \tag{43}$$

where $\lambda_{r_i^{\text{Wup}}}$ is the upward reserve demand price for wind power unit i; $\lambda_{\text{rup},i}^{\text{EN}}$ is the reserve demand's energy component; $\lambda_{\text{rup},i}^{\text{CO}}$ is the reserve demand's congestion component; $\xi_{w,i}$ represents the ratio of the wind power forecast error's standard deviation to its ancillary service demand; $\rho_{w,i,j}$ is the correlation coefficient between wind power unit i and the thermal power unit. The composition of the price for downward reserve demand $\lambda_{r_i^{\text{Wdn}}}$ in wind power is consistent with that of upward reserve demand.

In the ancillary service market, the marginal congestion cost reflects the extra system costs due to transmission capacity constraints for reserves; the reserve scarcity price reflects the scarcity of reserve resources provided by thermal power units, which is determined by the reserve service cost declared by the winning thermal power unit.

The energy component of reserve demand refers to the cost that wind power must pay for generation due to its uncertainty. Due to forecast errors in wind power output, flexible resources like thermal power units need to provide reserve generation capacity to maintain power balance, so wind power must cover the cost of this reserve capacity. The congestion component of reserve demand represents the transmission cost wind power must pay due to its random variability. To ensure reserve capacity for renewable energy, transmission lines must retain a certain margin to address power fluctuations caused by wind power variability, requiring wind power to pay for this transmission margin.

If there is no uncertainty in wind power, such as $\xi_{w,i} = 0$, both the energy and congestion components of its reserve demand are zero, meaning the wind power's ancillary service demand is also zero. As the wind power reduces its uncertainty level, the energy and congestion components of reserve demand decrease accordingly, thereby reducing the price of ancillary services, lessening the ancillary service costs it bears, and enhancing its total generation revenue.

The settlement for each market participant in the ancillary service market is as follows:

(1) Revenue for thermal power units

Constructing the Lagrangian function as shown in Appendix A.2 and taking the partial derivatives with respect to upward and downward spinning reserve capacity, the corresponding prices can be expressed as follows:

$$\lambda_{R,i}^{\mathrm{Gup}} = -\partial L / \partial r_i^{\mathrm{Gup}} = \eta_{R,i}^{\mathrm{Gup}} \tag{44}$$

$$\lambda_{R,i}^{\mathrm{Gdn}} = -\partial L / \partial r_i^{\mathrm{Gdn}} = \eta_{R,i}^{\mathrm{Gdn}} \tag{45}$$

$$R_{G_i}^{flex} = \eta_{R,i}^{\mathrm{Gup}} r_i^{\mathrm{Gup}} + \eta_{R,i}^{\mathrm{Gdn}} r_i^{\mathrm{Gdn}} \tag{46}$$

where $R_{G_i}^{flex}$ represents the revenue earned by the thermal power unit for providing reserve generation capacity.

(2) Payment by wind power units

$$PAY_{w,i} = \lambda_{r_i^{\mathrm{Wup}}} r_i^{\mathrm{Wup}} + \lambda_{r_i^{\mathrm{Wdn}}} r_i^{\mathrm{Wdn}} \tag{47}$$

where $PAY_{w,i}$ represents the ancillary service cost borne by wind power unit i.

The ancillary service costs borne by wind power units are calculated by multiplying their reserve demand prices by the declared reserve demand. The ancillary service allocation method proposed in this paper effectively differentiates the ancillary service demand of various renewable energy sources and establishes differentiated reserve demand prices based on the uncertainty severity inherent in each energy source. In traditional pricing methods, renewable energy units with higher uncertainty typically determine the marginal unit, thereby driving up the marginal reserve price. In contrast, the pricing mechanism proposed herein avoids the uniform pricing of renewable energy units with different uncertainty tiers, thereby preventing units with lower uncertainty from bearing excessive ancillary service costs. This differentiated pricing approach ensures that each unit bears ancillary service costs in proportion to its uncertainty level, incentivizing renewable energy units to better mitigate forecast variability, thus enhancing both the efficiency and fairness of the overall system. Furthermore, this mechanism aligns with the principles of incentive-compatible market design, fostering a fair distribution of costs among renewable energy sources and improving the rationality and transparency of ancillary service cost allocation.

The total generation revenue $R_{\mathrm{Wind},i}$ for wind power unit i is the difference between its energy market revenue and ancillary service market costs.

$$R_{\mathrm{Wind},i} = F_{\mathrm{Wind},i} - PAY_{\mathrm{Wind},i} \tag{48}$$

Thus, the unit generation price for wind power can be obtained as follows:

$$\lambda_{\text{av}} = R_{\text{Wind},i} / P_i^{\text{W}} \tag{49}$$

When the planned generation power of wind power is the same, wind power with higher variability will face greater ancillary service demand, resulting in a higher unit price for ancillary service demand. This leads to increased ancillary service costs, thereby reducing total generation revenue and lowering the unit generation price. Providing clear price signals helps incentivize renewable energy to actively reduce its variability, thus lowering reserve demand.

(3) Reserved line congestion surplus

$$F^{\text{RLCS}} = \sum_{l \in I} f_l \left(\overline{\mu}_l^{\text{ru}} - \underline{\mu}_l^{\text{ru}} + \overline{\mu}_l^{\text{rd}} - \underline{\mu}_l^{\text{rd}} \right) \tag{50}$$

Transmission lines generate market revenue through congestion surplus. The reserve transmission capacity constraint ensures sufficient transmission capacity is allocated in grid dispatch to guarantee the reliable delivery of reserve capacity, which is ultimately reflected in energy prices. In energy market settlements, power consumers bear the congestion surplus caused by reserve transmission capacity. Following the "cost causation" principle, renewable energy units requesting reserve capacity should compensate power consumers for this cost to ensure fair allocation and prevent market distortions.

With the ancillary cost redistribution to renewable energy, a portion of these payments is compensated to consumers as reserve line congestion surplus. This is due to the fact that power consumers bear the total congestion surplus of the lines in the energy market; therefore, according to the 'cost causation' principle, the congestion surplus on reserved lines should be borne by renewable energy itself, requiring renewable energy to compensate power users. The settlements for each market entity in the electric energy market and ancillary service market adhere to the principle of revenue and expenditure balance.

6. Case Study

6.1. System Parameters

The effectiveness of the pricing method proposed in this paper is validated using an improved IEEE 30-bus system, as shown in Figure 3. The system includes six thermal power units and six wind power units, with transmission power limits of 135 MW and 300 MW on branches 19 and 31, respectively, while other branches have no transmission power limits. The planned output and uncertainty information of wind power units are provided in Table 2, with a tolerable risk level of 5% set for chance constraints. The model established in this paper is solved iteratively using the CPLEX solver in the YALMIP optimization toolbox in MATLAB R2022a.

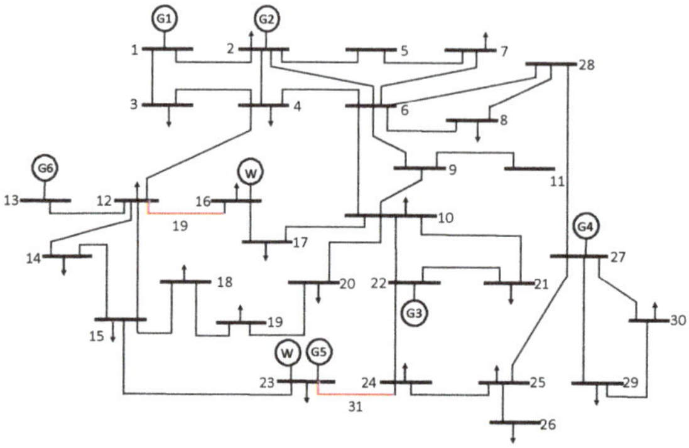

Figure 3. Improved topology of the IEEE30 node system.

Table 2. Pricing results for wind power unit reserve ancillary service demand in IEEE30 node System.

Wind Power Unit	Bus	Planned/MW	Mean/MW	Standard Deviation/MW
WF1	16	100	0	0
WF2	16	100	−2	10
WF3	16	100	−2	20
WF4	23	100	0	0
WF5	23	100	2	10
WF6	23	100	2	20

To verify the effectiveness of the pricing method proposed in this paper, the following two methods are used for comparison:

S1: Traditional uniform marginal pricing method [11].

S2: The differentiated pricing method considering uncertainty.

6.2. Clearing Results

Through joint clearing of the energy and ancillary service markets using the proposed distributionally robust optimization framework, the pricing results of wind power ancillary service demand under different pricing mechanisms are shown in Table 3, while the corresponding allocation of ancillary service costs among wind power units with varying uncertainty characteristics is systematically illustrated in Figure 4.

Table 3. Pricing results for wind power unit reserve ancillary service demand in IEEE30 node System.

	Price (USD/MWh)	Wind Power Unit				
		WF1/WF4	WF2	WF3	WF5	WF6
S1	$\lambda_{r_i}^{\mathrm{Wup}}$	20.75	20.75	20.75	20.75	20.75
	$\lambda_{r_i}^{\mathrm{Wdn}}$	10.37	10.37	10.37	10.37	10.37
S2	$\lambda_{\mathrm{rup},i}^{\mathrm{EN}}$	0	7.74	16.36	3.80	7.24
	$\lambda_{\mathrm{rup},i}^{\mathrm{CO}}$	0	1.83	3.87	7.50	14.32
	$\lambda_{r_i}^{\mathrm{Wup}}$	0	9.57	20.23	11.30	21.56
	$\lambda_{\mathrm{rdn},i}^{\mathrm{EN}}$	0	2.67	5.12	5.47	11.57
	$\lambda_{\mathrm{rdn},i}^{\mathrm{CO}}$	0	0	0	0	0
	$\lambda_{r_i}^{\mathrm{Wdn}}$	0	2.67	5.12	5.47	11.57

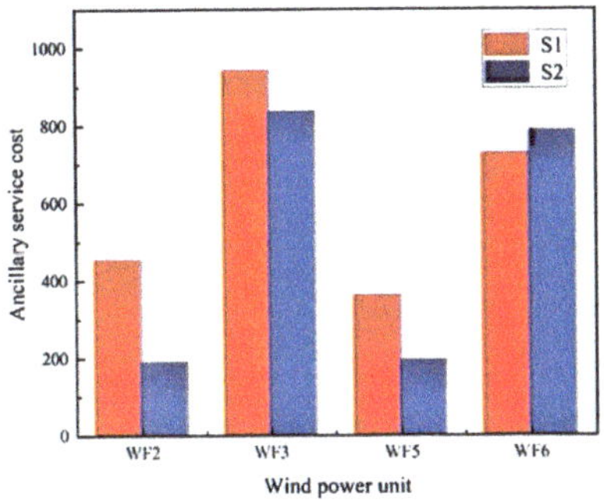

Figure 4. Allocation results of wind power units' ancillary service cost under different pricing mechanisms.

As observed from the table, under the traditional uniform marginal pricing mechanism, the upward and downward reserve demand prices for all wind power units are identical and determined by the marginal unit providing reserves. In contrast, under the differentiated pricing mechanism that considers uncertainty, the ancillary service demand price for wind power units consists of both energy and congestion components. A comparison of WF1, WF2, and WF3 with WF4, WF5, and WF6 reveals that, at the same node, the greater the uncertainty of a wind power unit, the higher its energy and congestion components for reserves, leading to an increase in the unit's ancillary service demand price. Notably, when the wind power output exhibits no fluctuations, the ancillary service demand price of that unit is zero. Taking WF3 and WF6 as examples, although both have the same standard deviation of forecast errors, WF3 has a negative mean forecast error, whereas WF6 has a positive mean forecast error. Consequently, WF3 has a greater upward reserve demand, while WF6 exhibits a more significant downward reserve demand. Accordingly, WF3 has a higher energy component for upward reserves and a lower energy component for downward reserves compared to WF6. Additionally, since WF6 is closer to the congested transmission line L31, its congestion component for upward reserves is relatively higher.

As shown in Figure 4, the allocation of ancillary service costs for wind power units differs significantly under different pricing mechanisms. The traditional uniform marginal pricing mechanism (S1) does not differentiate based on the uncertainty levels of wind power units, resulting in lower-uncertainty units (e.g., WF2 and WF5) bearing higher costs, while higher-uncertainty units (e.g., WF6) bear relatively lower costs. In contrast, the uncertainty-sensitive differentiated pricing mechanism (S2) aligns cost allocation with the uncertainty levels of wind power units, ensuring that units with higher uncertainty bear greater costs, while those with lower uncertainty see a cost reduction. This mechanism better adheres to market incentive principles, encouraging wind power units to reduce uncertainty and improving grid stability.

The pricing mechanism proposed in this paper effectively characterizes wind power uncertainty, accurately reflects network congestion levels and ancillary service supply–demand relationships, and provides clear price signals. Compared to the traditional uniform marginal pricing method, this mechanism ensures that wind power units with higher uncertainty bear higher ancillary service demand costs, thereby effectively mitigating the "free-rider" problem. Moreover, it incentivizes wind power units to proactively reduce their uncertainty and enhances the stability of system operations. This mechanism overcomes the limitations of traditional ancillary service pricing models by assessing the value of flexibility resources from the perspective of renewable energy as a demand-side participant. It captures the differences in flexibility demand across various renewable energy sources, establishing a refined and differentiated pricing framework. Ultimately, it provides theoretical support and practical guidance for building a more equitable and efficient electricity market.

The settlement of wind power in the energy and ancillary service markets is shown in Table 4.

Table 4. Pricing results for wind power unit reserve ancillary service demand in IEEE118 node System.

Wind Power Unit	Energy Revenue (USD)	Ancillary Service Cost (USD)	Total Revenue (USD)	Unit Generation Price (USD/MW)
WF1	1729.87	0	1729.87	17.30
WF2	1729.87	191.75	1538.12	15.38
WF3	1729.87	837.38	892.49	8.92
WF4	2963.14	0	2963.14	29.63
WF5	2963.14	194.02	2769.12	27.69
WF6	2963.14	785.78	2177.36	21.77

According to Table 4, the energy revenue for wind power units at the same node remains the same, but the cost of ancillary services increases with the rise in the wind power units' uncertainty. Consequently, the total generation revenue decreases, and the unit price of generation also drops. Considering the impact of uncertainty of renewable energy on actual system operation, differentiated cost-sharing is applied, whereby renewable energy with higher uncertainty bears a larger share of reserve costs, aligning with the principle of cost causation.

The settlement results for each market entity in the energy and ancillary service markets are presented in Table 5.

Table 5. Wind power units' settlement results in the energy and ancillary service markets.

Market Participation	Market Entities	Revenue and Expenditure Types	Settlement (USD)
Energy market	Thermal power unit	Energy revenue	14,455.88
	Wind power unit	Energy revenue	14,079.01
	Power consumer	Energy expenditure	−38,388.46
	Transmission line	Congestion surplus	9853.57
Ancillary service market	Thermal power unit	Upward reserve revenue	897.3
		Downward reserve revenue	634.34
	Wind power unit	Upward reserve payment	−1374.58
		Downward reserve payment	−634.34
	Power consumer	Reserved line compensation	477.28

Table 5 presents the settlement results for wind power units in both the energy and ancillary service markets, detailing the revenues and expenditures of various market entities. In the energy market, thermal power units and wind power units generate revenue from energy sales, as indicated by positive settlement values of USD 14,455.88 and USD 14,079.01, respectively. Power consumers incur an energy expenditure of USD −38,388.46, representing the cost of purchasing electricity. Additionally, transmission lines accumulate a congestion surplus of USD 9853.57. In the ancillary service market, wind power units pay upward and downward reserves based on their declared ancillary service demand (USD 1374.58 and USD 634.34, respectively), with a portion used to compensate thermal units that provide upward and downward reserves (USD 897.3 and USD 634.34, respectively) and a portion returned to power consumers as reserved line congestion surplus (USD 447.28). This is due to the fact that power consumers bear the total congestion surplus of the lines in the energy market; therefore, according to the "cost causation" principle, the congestion surplus on reserved lines should be borne by renewable energy itself, requiring renewable energy to compensate power users. The settlements for each market entity in the electric energy market and ancillary service market adhere to the principle of revenue and expenditure balance.

6.3. Sensitivity Analysis

The uncertainty severity in renewable energy determines the necessary ancillary service and reserved line margin, thereby affecting the clearing results and electricity price in the energy-ancillary service market. This section conducts a sensitivity analysis on the forecast error of wind power unit WF3. The up and down reserve demand prices and unit generation prices for WF3 are shown in Figure 5. The horizontal axis represents the scaling factor of the standard deviation of WF3's forecast error.

As shown in Figure 5, at a forecast error scaling factor of 0.2, the upward reserve demand price starts at 4 USD/MWh, the downward reserve at 2 USD/MWh, and the generation price at 17 USD/MWh. When the scaling factor reaches 1.2, the upward reserve increases 475% to 23 USD/MWh, the downward reserve rises 200% to 6 USD/MWh, while the generation price decreases 76% to 4 USD/MWh. As forecast error increases, the required reserve capacity and associated costs escalate, leading to higher upward reserve demand prices, which demonstrate a strong positive correlation between uncertainty and reserve prices, effectively allocating costs to other units. In addition, as the uncertainty of WF3 increases, both the up-and-down reserve demand prices for this unit gradually increase, while the unit generation price gradually decreases.

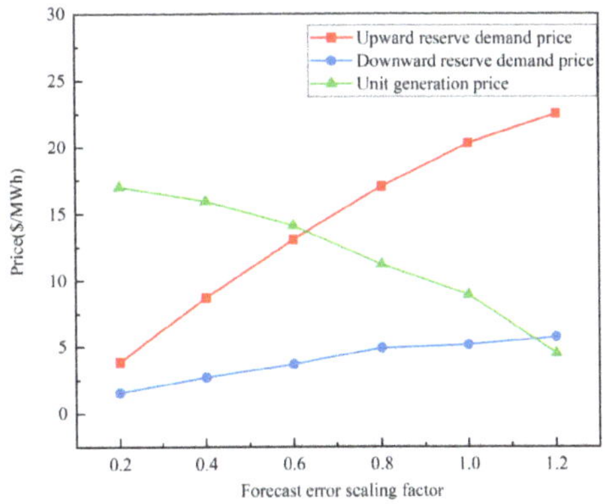

Figure 5. Impact of renewable energy uncertainty levels on pricing outcomes.

Figure 6 indicates that, as the scaling factor of WF3's forecast error increases, the total reserve service cost for the system and the reserve cost allocated to WF3 both gradually rise, while the reserve costs allocated to other wind power units decrease. As WF3's forecast error scaling factor increases from 0.2 to 1.2, WF3's allocated ancillary service cost escalates from approximately 250 USD/MWh to 1200 USD/MWh, while total system cost rises from 1400 USD/MWh to 2100 USD/MWh (50% increase). Other units' costs remain stable around 800–900 USD/MWh.

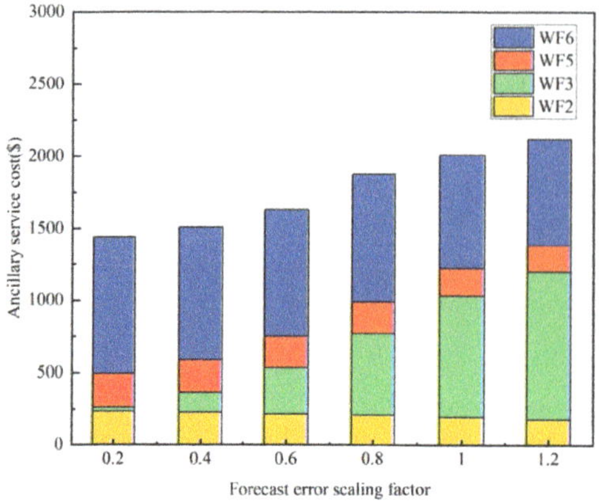

Figure 6. Impact of changes in individual renewable energy forecast errors on the cost of ancillary service.

The results reveal two critical mechanisms: (1) Systemic Impact: Individual unit uncertainty escalation drives both personal and system-wide cost increases, demonstrating localized uncertainty's broader effects. (2) Competitive Dynamics: Precise cost attribution ensures uncertainty reduction directly benefits responsible units without penalizing others, creating market-driven incentives for forecast accuracy improvements and advanced technology adoption across renewable energy sectors.

Additionally, a competitive mechanism among renewable energy plants emerges: when an individual plant reduces forecast error by enhancing the forecast accuracy, the ancillary service costs allocated to other plants increase.

6.4. Impact Analysis of Energy Storage Allocation

With the allocation of energy storage, the charging and discharging actions of the storage system can effectively regulate power output, reducing power deviations. This section focuses on the impact of unit energy storage on renewable energy forecast error and ancillary service demand price. Using wind power unit WF3 as an example, with the allocation of 1 MW and 2 MW storage units, the effects on ancillary service demand price, ancillary service cost, and total generation revenue are examined, as shown in Figure 7.

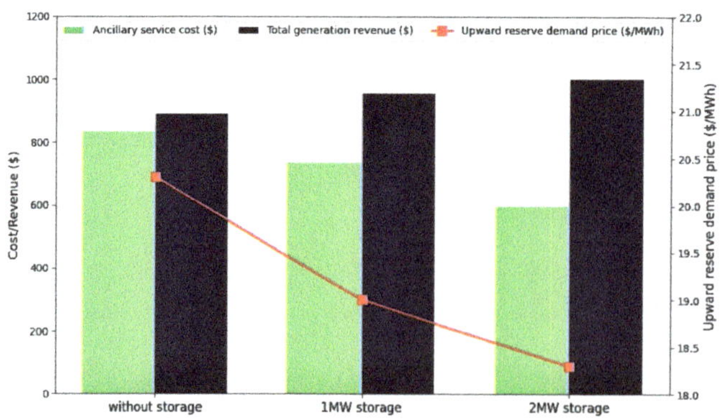

Figure 7. Changes in renewable energy reserve demand prices, ancillary service cost, and total revenue after the deployment of energy storage.

Figure 7 shows that after configuring energy storage, the price of up-reserve demand has decreased. The ancillary service cost has been reduced by 12.90% with 1 MW storage and further decreased by an additional 8.33% with 2 MW storage, while the total power generation revenue has increased by 6.73% with 1 MW storage and shows a further 4.71% improvement with 2 MW storage. This indicates that the integration of energy storage with renewable energy can optimize operations and improve profitability. The energy storage system's rapid response to power supply–demand changes reduces dependence on traditional flexible resources, thereby decreasing the demand for ancillary services. Furthermore, combining energy storage with renewable energy enhances renewable energy grid integration, allowing the power system to achieve greater flexibility and resilience, thus supporting the implementation of more renewable energy projects. This positive feedback effect has significant implications for advancing the energy transition.

6.5. IEEE-118 Bus System

To further validate the effectiveness of the proposed differentiated pricing method in large-scale systems, the standard IEEE 118-bus system is employed for verification. The system comprises 118 buses, 54 thermal power units, 186 transmission lines, and 20 wind power units. The tolerable risk level for chance constraints is set at 5%. In total, 20 wind power units are deployed across 6 key buses, with 2–4 units at each bus. These wind power units have different uncertainty levels. All wind power units are configured with a planned output of 100 MW to provide a clearer comparison of the impacts of different uncertainty levels. The planned output and forecasting error parameters of the 20 wind power units are shown in Table 6.

Table 6. Pricing results for wind power unit reserve ancillary service demand in IEEE118 node System.

Wind Power Unit	Bus	Planned/MW	Mean/MW	Standard Deviation/MW
WF1	16	100	0	0
WF2	16	100	−2	10
WF3	16	100	−2	20
WF4	23	100	0	0
WF5	23	100	2	10
WF6	23	100	2	20
WF7	23	100	0	0
WF8	49	100	−1	10
WF9	49	100	1	20
WF10	49	100	−3	30
WF11	49	100	0	0
WF12	65	100	1	10
WF13	65	100	−2	20
WF14	65	100	0	0
WF15	80	100	2	10
WF16	80	100	−1	20
WF17	80	100	3	30
WF18	100	100	0	0
WF19	100	100	−1	10
WF20	100	100	2	20

The settlement of wind power in the energy and ancillary service markets is shown in Table 7.

Table 7. The settlement and pricing results for the wind power unit reserve ancillary service demand.

Wind Power Unit	Energy Revenue (USD)	Ancillary Service Cost (USD)	Total Revenue (USD)	Unit Generation Price (USD/MW)
WF1	2156.34	0	2156.34	21.56
WF2	2156.34	178.45	1977.89	19.78
WF3	2156.34	356.90	1799.44	17.99
WF4	2834.72	0	2834.72	28.35
WF5	2834.72	198.23	2636.49	26.36
WF6	2834.72	396.46	2438.26	24.38
WF7	2834.72	0	2834.72	28.35
WF8	1947.58	167.89	1779.69	17.80
WF9	1947.58	335.78	1611.80	16.12
WF10	1947.58	503.67	1443.91	14.44
WF11	1947.58	0	1947.58	19.48
WF12	2278.91	189.34	2089.57	20.90
WF13	2278.91	378.68	1900.23	19.00
WF14	2278.91	0	2278.91	22.79
WF15	2512.43	203.67	2308.76	23.09
WF16	2512.43	407.34	2105.09	21.05
WF17	2512.43	611.01	1901.42	19.01
WF18	2687.92	0	2687.92	26.88
WF19	2687.92	195.78	2492.14	24.92
WF20	2687.92	391.56	2296.36	22.96

The settlement results demonstrate significant price dispersion among units with varying uncertainty levels at the same bus. Zero-volatility units achieve the highest energy price of up to USD 28.35/MW, while extreme volatility units receive as low as USD 14.44/MW, with a maximum price differential of USD 13.91/MW. Nodal position affects energy revenue, but uncertainty level determines ancillary service costs. The differentiated pricing mechanism effectively incentivizes wind power units to improve forecasting accuracy and reduce system operational risks.

7. Conclusions

The large-scale integration of renewable energy into the grid has introduced significant uncertainty to power systems, increasing the demand for ancillary services. To address the "free-rider" issue in existing pricing mechanisms, this paper develops a pricing mechanism for ancillary service demand based on the variability of renewable energy sources. A joint clearing model for energy and ancillary services is proposed, introducing chance constraints for reserve transmission line allocation and generation reserves, which fully account for the stochastic characteristics of renewable energy and effectively distribute ancillary service costs to responsible parties. The main findings are as follows:

(1) The proposed pricing method analytically represents the uncertainty of renewable energy, enabling differentiated pricing for ancillary service demand. Renewable units with higher volatility incur higher ancillary service costs per unit, providing clear price signals to market participants. This mechanism quantitatively attributes the external costs of varying renewable energy volatility, ensuring fair cost-sharing of ancillary services among renewable energy, while fulfilling properties of incentive compatibility and budget balance.

(2) Renewable units bearing higher ancillary service costs can mitigate their demand for these services and increase overall revenue by improving forecasting accuracy or installing local energy storage to smooth output, thereby actively reducing uncertainty and enhancing renewable energy integration. Simulation results on the modified IEEE 30-bus system demonstrate that energy storage units could dynamically balance renewable power

output, reducing ancillary service costs by 12.9% while increasing total generation revenue by 6.73%.

It should be noted that the proposed model has certain limitations: (1) The system model proposed in this paper adopts DC power flow, which is usually applied in transmission networks rather than distribution networks. For the distribution network, the proposed model is not suitable. (2) The load forecast error is omitted, which is usually applied when the load forecast accuracy is high and the load uncertainty can be ignored. For the scenario where the load uncertainty is high, the proposed model is not suitable. In addition, the proposed electricity market pricing mechanism primarily targets the day-ahead market. Future research will focus on implementing pricing and settlement for renewable energy in the real-time market, along with establishing corresponding regulatory mechanisms.

Author Contributions: Conceptualization, X.L.; Methodology, X.Y.; Software, Q.A.; Validation, J.L. All authors have read and agreed to the published version of the manuscript.

Funding: This work was supported by the National Natural Science Foundation of China (No. 52277092).

Data Availability Statement: Data is contained within the article.

Conflicts of Interest: The authors declare no conflict of interest.

Appendix A

Appendix A.1. Analytical Transformation of Chance Constraints

The upward and downward reserve demands of renewable energy will alter the line power flow. Considering the deviation balance adjustment constraint, the branch power accounting for the upward reserve demand can be expressed as follows:

$$PF_l = f_l - \sum_{i \in I_W} \left(PTDF_{l-i} - \sum_{m \in I_G} PTDF_{l-m}\beta_m \right) r_i^{\mathrm{Wup}} \tag{A1}$$

The sensitivity of branch power to the upward reserve demand for wind power can be expressed as follows:

$$K_{\mathrm{PF,up},li} = \frac{\partial PF_l}{\partial r_i^{\mathrm{Wup}}} = -PTDF_{l-i} + \sum_{m \in I_G} PTDF_{l-m}\beta_m \tag{A2}$$

Equation (8) can be analytically transformed into the following:

$$f_l + \sum_{i \in I_W} K_{\mathrm{PF,up},li}\mu_i + \kappa_\varepsilon \sqrt{{K_{\mathrm{PF,up}}}^T \Sigma_{\mathrm{W}} K_{\mathrm{PF,up}}} \le F_l \tag{A3}$$

where μ_i is the mean of the forecasting errors of renewable energy, and Σ_{W} is the covariance matrix of the forecasting errors of renewable energy.

Similarly, Equations (9)–(11) can be analytically transformed into the following:

$$f_l + \sum_{i \in I_W} K_{\mathrm{PF,up},li}\mu_i - \kappa_\varepsilon \sqrt{{K_{\mathrm{PF,up}}}^T \Sigma_{\mathrm{W}} K_{\mathrm{PF,up}}} \ge -F_l \tag{A4}$$

$$f_l + \sum_{i \in I_W} K_{\mathrm{PF,dn},li}\mu_i + \kappa_\varepsilon \sqrt{{K_{\mathrm{PF,dn}}}^T \Sigma_{\mathrm{W}} K_{\mathrm{PF,dn}}} \le F_l \tag{A5}$$

$$f_l + \sum_{i \in I_W} K_{PF,dn,li}\mu_i - \kappa_\varepsilon \sqrt{K_{PF,dn}{}^T \Sigma_W K_{PF,dn}} \geq -F_l \tag{A6}$$

where $K_{PF,dn,li}$ represents the sensitivity of branch power to the downward reserve demand for wind power.

The upward and downward reserve demands of renewable energy will also alter the upward and downward spinning reserve capacity provided by renewable sources. The sensitivity of the upward reserve provided by thermal power units to the upward reserve demand for wind power is given by the following:

$$K_{RG,up,ij} = \frac{\partial R_{Gi}}{\partial r_j^{Wup}} = \beta_i \tag{A7}$$

Equation (18) can be analytically transformed into the following:

$$-\beta_i \sum_{j \in I_W} \mu_j + \kappa_\varepsilon \sqrt{K_{RG,up}{}^T \Sigma_W K_{RG,up}} \leq r_i^{Gup} \tag{A8}$$

Similarly, Equation (19) can be transformed into

$$-\gamma_i \sum_{j \in I_W} \mu_j + \kappa_\varepsilon \sqrt{K_{RG,dn}{}^T \Sigma_W K_{RG,dn}} \leq r_i^{Gdn}, \tag{A9}$$

where $K_{RG,dn}$ is the sensitivity matrix of the downward reserve provided by thermal power units to the downward reserve demand for wind power.

Thus, the nonlinear chance-constrained economic dispatch model proposed above is approximated as a second-order cone optimization problem, which can be solved directly using commercial solvers.

Appendix A.2. Derivation of Reserve Demand Pricing

Taking the analytical model mentioned in the previous section as an example, its Lagrangian function is shown as follows:

$$\begin{aligned}
L = & \sum_{i \in I_G} \left(a_i (P_i^G)^2 + b_i P_i^G + c_i \right) + \sum_{i \in I_G} \left(m_i^{up} r_i^{Gup} + m_i^{dn} r_i^{Gdn} \right) \\
& - \lambda_e \left(\sum_{i \in I_G} P_i^G + \sum_{i \in I_W} P_i^W - \sum_{i \in I_D} D_i \right) - \sum_{l \in I} \mu_l^{max} (-f_l + F_l) - \sum_{l \in I} \mu_l^{min} (f_l + F_l) \\
& - \overline{\mu}_l^{ru} \left(-f_l - \sum_{i \in I_W} K_{PF,up,li}\mu_{up,i} - \kappa_\varepsilon \sqrt{K_{PF,up}{}^T \Sigma_W K_{PF,up}} + F_l \right) - \underline{\mu}_l^{ru} \left(f_l + \sum_{i \in I_W} K_{PF,up,li}\mu_{up,i} - \kappa_\varepsilon \sqrt{K_{PF,up}{}^T \Sigma_W K_{PF,up}} + F_l \right) \\
& - \overline{\mu}_l^{rd} \left(-f_l - \sum_{i \in I_W} K_{PF,dn,li}\mu_{dn,i} - \kappa_\varepsilon \sqrt{K_{PF,dn}{}^T \Sigma_W K_{PF,dn}} + F_l \right) - \underline{\mu}_l^{rd} \left(f_l + \sum_{i \in I_W} K_{PF,dn,li}\mu_{dn,i} - \kappa_\varepsilon \sqrt{K_{PF,dn}{}^T \Sigma_W K_{PF,dn}} + F_l \right) \\
& - \sum_{i \in I_G} \eta_{R,i}^{Gup} \left(\beta_i \sum_{j \in I_W} \mu_{up,j} - \kappa_\varepsilon \sqrt{K_{RG,up}{}^T \Sigma_W K_{RG,up}} + r_i^{Gup} \right) - \sum_{i \in I_G} \eta_{R,i}^{Gdn} \left(\gamma_i \sum_{j \in I_W} \mu_{dn,j} - \kappa_\varepsilon \sqrt{K_{RG,dn}{}^T \Sigma_W K_{RG,dn}} + r_i^{Gdn} \right)
\end{aligned} \tag{A10}$$

Taking the partial derivative with respect to the reserve demand of the wind power units yields the upward and downward reserve demand prices for wind power.

$$\lambda_{r_i^{\mathrm{Wup}}} = \frac{\partial L}{\partial r_i^{\mathrm{Wup}}} = \sum_{l \in I} \kappa_\varepsilon (\overline{\mu}_l^{\mathrm{ru}} + \underline{\mu}_l^{\mathrm{ru}}) \frac{K_{\mathrm{PF,up},li}{}^2 \xi_{w,i}{}^2 r_i^{\mathrm{Wup}} + K_{\mathrm{PF,up},li} \xi_{w,i} \sum_{j \in I_{\mathrm{W}}} \rho_{w,i,j} K_{\mathrm{PF,up},lj} \xi_{w,j} r_i^{\mathrm{Wup}}}{\sqrt{K_{\mathrm{PF,up}}{}^T \Sigma_{\mathrm{W}} K_{\mathrm{PF,up}}}} + \sum_{i \in I_{\mathrm{G}}} \kappa_\varepsilon \eta_{\mathrm{R},i}^{\mathrm{Gup}} \frac{\beta_i^2 \xi_{w,i}{}^2 r_i^{\mathrm{Wup}} + \beta_i \xi_{w,i} \sum_{j \in I_{\mathrm{W}}} \rho_{w,i,j} \beta_{w,j} \xi_j r_i^{\mathrm{Wup}}}{\sqrt{K_{\mathrm{RG,up}}{}^T \Sigma_{\mathrm{W}} K_{\mathrm{RG,up}}}} \tag{A11}$$

$$\lambda_{r_i^{\mathrm{Wdn}}} = \frac{\partial L}{\partial r_i^{\mathrm{Wdn}}} = \sum_{l \in I} \kappa_\varepsilon (\overline{\mu}_l^{\mathrm{rd}} + \underline{\mu}_l^{\mathrm{rd}}) \frac{K_{\mathrm{PF,dn},li}{}^2 \xi_{w,i}{}^2 r_i^{\mathrm{Wdn}} + K_{\mathrm{PF,dn},li} \xi_{w,i} \sum_{j \in I_{\mathrm{W}}} \rho_{w,i,j} K_{\mathrm{PF,dn},lj} \xi_{w,j} r_i^{\mathrm{Wdn}}}{\sqrt{K_{\mathrm{PF,dn}}{}^T \Sigma_{\mathrm{W}} K_{\mathrm{PF,dn}}}} + \sum_{i \in I_{\mathrm{G}}} \kappa_\varepsilon \eta_{\mathrm{R},i}^{\mathrm{Gdn}} \frac{\gamma_i^2 \xi_{w,i}{}^2 r_i^{\mathrm{Wdn}} + \gamma_i \xi_{w,i} \sum_{j \in I_{\mathrm{W}}} \rho_{w,i,j} \gamma_{w,j} \xi_j r_i^{\mathrm{Wdn}}}{\sqrt{K_{\mathrm{RG,dn}}{}^T \Sigma_{\mathrm{W}} K_{\mathrm{RG,dn}}}} \tag{A12}$$

where $\lambda_{r_i^{\mathrm{Wup}}}$ and $\lambda_{r_i^{\mathrm{Wdn}}}$ are the upward and downward reserve demand prices for renewable energy, respectively. They consist of the reserve energy component and the blocking component, representing the costs that renewable energy units should pay for each unit of reserve demand.

References

1. Wei, Y. Study on Renewable Energy Participation Strategies in Electricity Market Trading in the Context of New Power System. *Res. Ind. Innov.* **2024**, 101–103.
2. Herding, R.; Ross, E.; Jones, W.R.; Endler, E.; Charitopoulos, V.M.; Papageorgiou, L.G. Risk-aware microgrid operation and participation in the day-ahead electricity market. *Adv. Appl. Energy* **2024**, *15*, 100180. [CrossRef]
3. National Development and Reform Commission. Available online: https://www.gov.cn/zhengce/zhengceku/2021-06/11/content_5617297.htm (accessed on 7 July 2021).
4. Qie, Y. The financial impact of the pricing mechanism of renewable energy power trading market. *Economist* **2025**, 64–65+68.
5. Figueiredo, N.C.; Silva, P.P. The "Merit-order effect" of wind and solar power: Volatility and determinants. *Renew. Sustain. Energy Rev.* **2019**, *102*, 54–62. [CrossRef]
6. Biber, A.; Felder, M.; Wieland, C.; Spliethoff, H. Negative price spiral caused by renewable energy? Electricity price pre-diction on the German market for 2030. *Electr. J.* **2022**, *35*, 107209.
7. Zhang, T.; Hu, Y.; Zhang, J.; Han, L.; Wang, P.; Chen, Y. Analysis of power market capacity remuneration mechanisms adapted to high penetration of renewable energy development. *Electr. Power Constr.* **2021**, *42*, 117–125.
8. Jiang, H.; Yue, C.; Yan, X. Influence of system inertia on flexibility resource analysis for an interconnection system with a high proportion of intermittent renewable energy. *Power Syst. Prot. Control* **2021**, *49*, 44–51.
9. Xia, Q.; Fang, Z.; Lai, X.; Zhang, Y.; Xiang, M.; Meng, F. Electricity Market Design Based on Temporal Pricing of Renewable Capacity. *Power Syst. Technol.* **2022**, *46*, 1771–1779.
10. Sun, D.; Shi, X.; Hai, S.; Xu, L.; Yang, Z.; Yang, C. Design of Electricity Ancillary Service Market System under the Environment of National Unified Electricity Market. *Autom. Electr. Power Syst.* **2024**, *48*, 13–24.
11. Liu, S.; Zhang, M.; Xia, Q. Design of Spot Market Mechanism Based on Demand Bidding of Reserve Ancillary Service. *Autom. Electr. Power Syst.* **2022**, *46*, 72–82.
12. Shu, K.; Sun, Z.; Huang, Y.; Li, R.; Wang, H.; Fang, P.; Luo, J. Research on Cost Allocation Methods for Frequency Regulation Auxiliary Services-Based on quantitative analysis of frequency regulation capacity and frequency regulation Mileage responsibility of power system grid-connected entities. *Price Theory Pract.* **2024**, 93–98.
13. Liu, Q.; Jiang, Y.; Zheng, Z.; Yang, S.; Deng, Y.; Li, M. Comparative Analysis on Electricity Market Balancing Mechanisms of Germany and UK and Its Enlightenment. *Autom. Electr. Power Syst.* **2024**, *48*, 8–15.
14. Bai, Y.; Dai, M.; Ji, T.; Jing, Z.; Zhong, J. Trading Mechanism of Spot Electric Energy and Frequency Regulation Markets Considering Renewable Energy Under the Background of New Power System. *South. Power Syst. Technol.* **2024**, *18*, 1–11.

15. Fang, X.; Du, E.; Zheng, K.; Yang, J.; Chen, Q. Locational pricing of uncertainty based on robust optimization. *CSEE J. Power Energy Syst.* **2021**, *7*, 1345–1356.
16. Xia, Y.; Xu, Q.; Huang, Y.; Du, P. Peer-to-peer energy trading market considering renewable energy uncertainty and participants individual preferences. *Int. J. Electr. Power Energy Syst.* **2023**, *148*, 108931. [CrossRef]
17. Brooks, A.E.; Lesieutre, B.C. A locational marginal price for frequency balancing operations in regulation markets. *Appl. Energy* **2022**, *308*, 118306. [CrossRef]
18. Lei, X.; Yang, Z. Electricity Market Pricing Method Considering Multiple Statistical Moments of Renewable Energy Uncertainty. In Proceedings of the CSEE, Lisbon, Portugal, 29–31 March 2023; Volume 43, pp. 5772–5785.
19. Wang, H.; Bie, Z.; Ye, H. Locational Marginal Pricing for Flexibility and Uncertainty with Moment Information. *IEEE Trans. Power Syst.* **2022**, *38*, 2761–2775. [CrossRef]
20. Xu, Z.; Li, J.; Liu, H.; Ge, S.; Gu, C.; Tian, W. A novel locational pricing approach for day-ahead distribution market under multiple uncertainties. *CSEE J. Power Energy Syst.* **2023**, 1–11.
21. Fu, Y.; Song, Y.; Deng, L.; Yang, L. Mechanism Design for Integrated Energy Market Considering Renewable Energy Uncertainties. *Power Syst. Technol.* **2024**, *49*, 1848–1858.

energies

MDPI

Article

Process Control via Electrical Impedance Tomography for Energy-Aware Industrial Systems

Krzysztof Król [1,2,*], Grzegorz Kłosowski [3], Tomasz Rymarczyk [1,2], Konrad Gauda [1], Monika Kulisz [3], Ewa Golec [4] and Agnieszka Surowiec [3]

1 WSEI University, Faculty of Transportation and Information Technology, Projektowa 4, 20-209 Lublin, Poland; tomasz@rymarczyk.com (T.R.); konrad.gauda@wsei.pl (K.G.)
2 Research and Development Center, Netrix S.A., 20-704 Lublin, Poland
3 Lublin University of Technology, Faculty of Management, Nadbystrzycka 38, 20-618 Lublin, Poland; g.klosowski@pollub.pl (G.K.); m.kulisz@pollub.pl (M.K.); a.surowiec@pollub.pl (A.S.)
4 WSEI University, Faculty of Administration and Social Sciences, Projektowa 4, 20-209 Lublin, Poland; ewa.golec@wsei.pl
* Correspondence: krzysztof.krol@wsei.pl

Abstract

Conventionally, tomography is an inspection technique in which tomographic images are intended for human perception and interpretation. In this work, we shift this paradigm by transforming tomography into an autonomous estimator of industrial reactor states, enabling fully automated process control. Alcoholic fermentation was employed as an example of a controlled process in the current study. The work presents an original concept utilizing transfer learning in conjunction with a ResNet-type artificial neural network, which converts electrical measurements into a sequence of values correlated with the conductivity of pixels constituting the cross-section of the examined biochemical reactor. The conductivity vector is transformed into a parameter determining substrate concentration, enabling dynamic process regulation in response to signals generated from EIT (Electrical Impedance Tomography). Within the scope of the described research, calibration of the conductivity vector against substrate concentrations was performed, and a Matlab/Simulink-based dynamic Monod kinetics model was developed. The obtained results demonstrate high accuracy in substrate concentration estimation relative to reference values throughout a forty-six-hour process. The same signals enable energy-efficient process control, in which cooling and mixing intensity are regulated according to energy prices and renewable energy availability. This strategy may possess particular application in facilities where fermentation installations are co-located with bioenergy production units.

Keywords: electrical impedance tomography (EIT); adaptive control; energy-aware industrial systems; neural networks

1. Introduction

1.1. Background and Related Work

Modern process plants in the chemical, food, pharmaceutical and water–wastewater sectors are moving from reactive control to proactive strategies under Quality by Design and Process Analytical Technology (PAT) frameworks [1,2]. The central challenge in continuous operations is to recover reliable, non-invasive and disturbance-resistant information about the internal state of equipment under flowing streams, strong couplings, and hard operating constraints. Point sensors such as pH, conductivity, temperature and flow are valuable but inherently local, which is often insufficient for precise model-based control in the presence of heterogeneous fields [3,4]. Spectrophotometry (UV–Vis) is widely used to quantify key analytes via the Beer–Lambert relationship between absorbance, concentration, and optical path length. In practice, measurements are often performed off-line on extracted samples, although in-line fiber-optic implementations exist within PAT deployments [5]. By contrast, tomography provides spatially resolved information suitable for machine-readable state estimation at high refresh rates.

Process tomography addresses this gap by providing cross-sectional images of physical quantities that correlate with the composition and structure of the medium. Several modalities are used in practice, including electrical impedance tomography (EIT) [6,7], electrical capacitance tomography (ECT) [7,8], ultrasound tomography (UST) [9,10], nuclear magnetic resonance (NMR) or magnetic resonance imaging (MRI) [11], and optical coherence tomography (OCT) [12,13]. Among these, EIT stands out for its bioprocess safety, comparatively low hardware cost, high sampling rates and ease of integration with production lines. Its inverse problem is ill-posed and sensitive to contact artifacts, yet recent methods in image reconstruction and physics-aware data fusion have markedly reduced these limitations and improved readiness for online deployment [14–16]. In bioreactors, practical confounders such as electrode polarization and fouling, temperature and ionic strength drifts, and gas holdup from CO_2 must be addressed through design and compensation. Practical demonstrations in low conductivity systems, e.g., combining electrical impedance spectroscopy with 2D electrical resistance tomography to monitor antisolvent crystallization, further illustrate the potential of impedance-based imaging for industrial monitoring [17].

Beyond diagnostic use, several studies show that tomographic measurements can support control decisions in particulate processes. In control-oriented frameworks, ECT images or derived features are used to regulate multiphase flows and fluidized beds, indicating practical pathways toward closed-loop operation in continuous settings [18]. At the same time, high-speed ECT systems have been demonstrated for online monitoring of pneumatic conveying with acquisition rates in the hundreds of frames per second, meeting the timing requirements of industrial supervision and control [19].

Biotechnological processes, including alcoholic fermentation, are a natural application for EIT. Bioethanol production is carried out in batch, semi continuous and continuous modes. Decision variables include dilution factor D, nutrient doses, temperature, early aeration, and cooling and mixing schedules. Substrate, biomass, and product concentrations are strongly correlated with solution conductivity and its distribution. Conductivity maps can therefore power soft sensors for hidden state estimation and subsequently for monitoring and predictive control. In recent years, machine learning based soft sensor solutions for fermentation have been presented, including variants designed and maintained according to Machine Learning Operations (MLOps) practices [20], as well as in-line implementations using Raman and Near Infrared Spectroscopy (NIR) spectroscopy [21], and electronic noses [22]. A similar trend is demonstrated by work combining impedance sensors with data models for ethanol and biomass monitoring [23].

Progress over the last few years shows that combining EIT with deep learning image reconstruction, and with non-structured kinetic models such as Monod relations with inhibitions, improves the stability and repeatability of process state estimation [24,25]. The resulting processing chain from EIT measurements through image reconstruction and feature extraction to a soft sensor, followed by an observer and controller, enables not only tight feedback but also supervisory control. In supervisory layers, schedules of energy-intensive auxiliaries such as cooling, recirculation and pumps can be coordinated with technological constraints and energy price or availability signals, which is increasingly relevant under high shares of renewables [14,15]. Related industrial studies also explore energy recovery options in auxiliary systems, e.g., compressed-air engines, underscoring the broader context for energy-aware operation [26]. These observations define a pathway in which tomographic sensing advances from diagnostics to an active component of control architectures for continuous bioprocessing. EIT provides spatially resolved fields that can be converted into stable numerical features and linked to process kinetics with latency compatible with real-time decision making. In this systems view, the primary objective is not image appearance but the stability, informativeness, and timeliness of features that support observers, feedback regulation, and energy-aware supervision under industrial constraints.

1.2. Motivation, Novelty and Paper Structure

The increasing penetration of renewable energy sources into power systems necessitates advanced monitoring and control strategies to ensure stability, efficiency, and resilience. Industrial process control faces similar challenges, as complex, nonlinear, and spatially distributed processes must be monitored and regulated in real time to meet quality, safety, and energy-efficiency requirements. In this context, process tomography has emerged as a promising tool for non-invasive visualization of internal states in biochemical tank reactors and pipelines. Electrical Impedance Tomography (EIT) is particularly attractive because of its low-cost instrumentation, high sampling rates, and ability to operate under industrial conditions without interrupting production.

While EIT has been extensively studied for monitoring and diagnostic purposes, its application to automatic, unattended process control remains largely unexplored. Existing literature has primarily focused on producing tomograms interpretable by human operators, emphasizing visual clarity and inclusion detectability. However, when EIT is used as part of a closed-loop control system, the human operator is no longer the primary recipient of the reconstructed images. Instead, the outputs must serve as numerical features that enable automated decision-making. In this paradigm, the visual quality of the reconstruction is secondary, as the controller consumes vectors of features or scalar indicators derived from tomograms to regulate the process in real time. This conceptual shift—from human-centered visualization to machine-readable state estimation—represents the key novelty of this work.

The primary objective of this research is to transform Electrical Impedance Tomography (EIT) from a diagnostic imaging tool into an autonomous process control sensor for biochemical systems, specifically addressing the gap between traditional human-interpreted tomographic visualization and automated industrial process control. This study aims to develop and validate an integrated framework that converts EIT measurements directly into machine-readable state vectors, eliminating the need for human interpretation of tomographic images in closed-loop control applications. The specific goals include (1) establishing a paradigm shift from visual tomogram interpretation to numerical feature extraction for automated decision-making, (2) integrating real-time EIT measurements with Monod kinetics modeling to enable accurate state estimation of fermentation processes, and (3) demonstrating energy-efficient process control by coordinating tomography-informed decisions with external energy constraints such as renewable energy availability and demand-response signals.

The proposed framework integrates EIT-based state reconstruction with machine learning models for feature extraction and prediction, followed by supervisory control algorithms that can adjust process variables such as dilution rate, temperature, and mixing schedules. This approach allows the coordination of process dynamics with external energy constraints, including demand-response signals and renewable energy availability, which is crucial for sustainable operation in modern distributed energy systems.

The remainder of the paper is organized as follows. Section 2 describes the fermentation process selected as a representative industrial case study and details the experimental setup and the EIT data acquisition system. The machine learning-based reconstruction and feature extraction pipeline, as well as the kinetic modeling approach, are also introduced. Section 3 presents simulation results demonstrating the closed-loop behavior of the system under repeated batch fermentation cycles. Section 4 discusses the implications of using EIT for process control, its advantages and limitations, and the potential for integration with energy-aware scheduling. Finally, Section 5 summarizes the main contributions and outlines future research directions, including extension to other industrial processes and coupling with advanced energy management frameworks.

2. Materials and Methods

2.1. Description of the Alcoholic Fermentation Process

Fermentation is observed in natural environments and is widely applied across numerous production domains. Depending on the specific pathway, various gases are released. It is defined as an anaerobic mode of respiration performed by many bacteria and certain fungi. During fermentation, the respiratory substrate is decomposed and transformed, with one product undergoing oxidation and another reduction. Energy is generated and stored as ATP (adenosine 5′-triphosphate). Based on the principal end product, lactic, butyric, acetic, alcoholic, and methanogenic fermentation are distinguished [27]. Alcoholic fermentation is considered here as the target process for sensing and control. It is defined as the anaerobic conversion of carbohydrates, predominantly by *Saccharomyces cerevisiae*, that yields ethanol and carbon dioxide. The process is widely used in the distilling, brewing, and wine industries. Its biological basis is the ability of yeast to hydrolyze and process simple sugars in the presence of appropriate enzymes.

The choice of alcoholic fermentation is motivated by practical considerations that support stable operation and reproducible experimentation. The course of alcoholic fermentation under typical conditions is predictable and relatively insensitive to moderate temperature variation, which facilitates controlled trials. From a monitoring perspective, the primary observable is the spatiotemporal evolution of bulk electrical conductivity inside the fermenter. Conductivity depends on the concentration of ions dissolved in the aqueous phase and on the accumulation of fermentation products. In alcoholic fermentation, fermentable sugars are initially present in solution and ethanol is a principal product with a characteristic electrical signature, so the conductivity of the medium changes in a gradual and trackable manner as sugars are converted to ethanol and carbon dioxide.

At the biochemical level, fermentable sugars such as glucose, fructose, and sucrose enter glycolysis to form pyruvate. Pyruvate is then decarboxylated to acetaldehyde and reduced to ethanol, and during this reduction, the oxidized form of nicotinamide adenine dinucleotide is regenerated, which maintains redox balance and allows glycolysis to continue under oxygen limitation [28]. A canonical overall stoichiometry on glucose is (1)

$$C_6H_{12}O_6 \rightarrow C_2H_5OH + 2CO_2, \tag{1}$$

which summarizes the conversion to ethanol and carbon dioxide with substrate-level phosphorylation. *Saccharomyces cerevisiae* is preferred in industrial practice because it combines rapid uptake and metabolism of sugars with tolerance to osmotic stress and elevated ethanol concentrations, supported by membrane remodeling and other stress response mechanisms [29,30]. The organism grows best in mildly acidic media and requires yeast assimilable nitrogen and micronutrients for enzyme synthesis and balanced growth, and the availability of these nutrients modulates both the fermentation rate and the formation of secondary metabolites [31].

The temporal evolution of the process begins with a short adaptation phase, after which active fermentation proceeds with rapid sugar depletion, biomass growth, and increasing formation of ethanol and carbon dioxide. As residual sugars decline and ethanol accumulates, the culture enters a decelerating phase that culminates in completion when sugars are exhausted or when inhibitory thresholds are reached. Product inhibition by ethanol, substrate inhibition at very high initial sugar concentrations, and nutrient limitations are the principal kinetic constraints. Operating conditions are typically mesophilic and mildly acidic, with agitation adjusted to maintain homogeneity and effective mass and heat transfer while avoiding excessive shear. Oxygen is minimized after the early adaptation period to favor fermentative metabolism, although brief controlled oxygen exposure at the start can support sterol and unsaturated fatty acid synthesis that stabilizes cell membranes. Alongside ethanol and carbon dioxide, minor compounds such as higher alcohols, organic acids, and esters are formed, and their concentrations depend on strain genetics, substrate composition, temperature, and nutrient status. Endpoint determination is based on the depletion of fermentable sugars to target levels and on the approach of carbon dioxide evolution and ethanol formation rates to quasi-steady values, subject to quality and safety specifications [32].

The conductivity $\rightarrow$ concentration mapping used in this research is an empirical, linear calibration fitted to alcoholic fermentation under our operating window, with a proportionality coefficient of 0.99 $\text{kg}\cdot\text{m}^{-3}$ per conductivity unit and an offset of -0.85 $\text{kg}\cdot\text{m}^{-3}$. This calibration is expected to transfer across batches with similar media, temperature, and ionic strength, but we treat it as site- and recipe-specific: different substrate compositions, yeast strains, or temperature profiles alter conductivity–chemistry coupling and typically require re-fitting the coefficients using a small set of paired EIT and reference measurements. Within a given facility, linearity has proven adequate over the relevant concentration range. If departures from linearity appear (e.g., due to high ethanol or strong ionic changes), we will adopt piecewise-linear or multivariable calibrations that incorporate temperature compensation and auxiliary probes while preserving the same estimation and control architecture.

2.2. Experimental Setup

The entire experimental stand, including the measurement vessel, the wireless sensor modules, and the electrical impedance tomography device, was manufactured by Netrix S.A. (Lublin, Poland). The experimental stand comprised a measurement vessel, a stirring system, and an integrated suite of sensors. The vessel was instrumented for electrical impedance tomography and operated as a bench-scale fermenter. Two circumferential rings of electrodes were mounted on the vessel wall in fixed positions in order to define a reproducible tomographic boundary, with the lower ring carrying electrodes one through sixteen and the upper ring carrying electrodes seventeen through thirty-two. Mixing was provided by a magnetic stirrer placed beneath the vessel, which avoided mechanical penetrations and mitigated flow-related artefacts in the vicinity of the electrode planes (Figure 1). Process pH was measured in the central region of the vessel using a laboratory pH meter (Figure 1b). Air temperature, relative humidity, and barometric pressure were monitored continuously during each run. Process temperature was measured at three locations to capture vertical gradients and boundary effects. One probe was located below the vessel, one above the vessel, and one within the liquid column between the two electrode rings.

(**a**) (**b**) (**c**)

Figure 1. Tank reactor with EIT electrodes and the magnetic stirrer: (**a**) adding substrate (sugar) to the vessel; (**b**) pH measurement; (**c**) controlled fermentation process in a closed vessel.

The principal process variable of interest was the electrical conductivity of the liquid phase. Conductivity depends on the concentration of dissolved ions and on the presence and evolution of fermentation products. During alcoholic fermentation, the progressive conversion of dissolved sugars to ethanol and carbon dioxide results in a gradual and measurable change in bulk conductivity, which provides an informative signature for subsequent analysis. The preparation of the medium followed a standard protocol. Three and a half liters of water were introduced into the vessel and one kilogram of sugar was added and mixed until fully dissolved. After complete dissolution, 23.37 g of dry distiller's yeast were added as inoculum. The vessel equipped with the electrode array was positioned on the magnetic stirrer and mixing was initiated. The preparations for the sugar solution and inoculation are documented in Figure 1a.

To perform measurements on a real physical model, a prototype EIT device designed and manufactured in the Netrix SA laboratory was used (Figure 2).

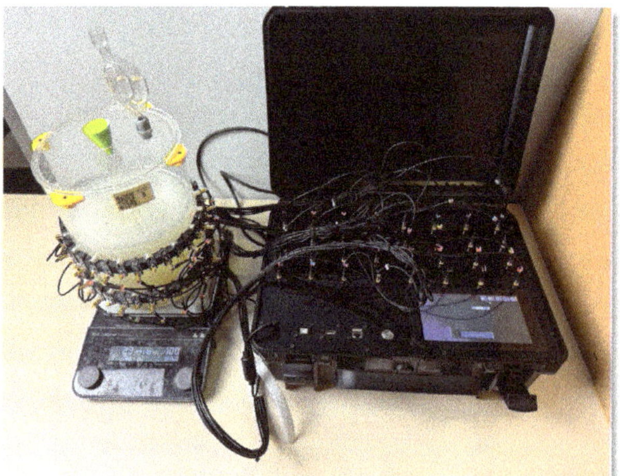

Figure 2. Measurement setup: prototype EIT device connected to the physical tank reactor model.

Data acquisition and sequencing were coordinated by the STMicroelectronics STM32F746G DISCO Discovery board (Agrate Brianza, Italy). The board is built around the STM32F746NGH6 microcontroller with an ARM Cortex M7 core clocked at 216 MHz. It provides 1 MB of embedded flash memory and 340 KB of static random access memory. Local human–machine interaction is supported by a color TFT LCD with a diagonal of 4.3 in, a resolution of 480 × 272 pixels, and a 16-bit color palette, together with a capacitive touch layer. A wired network interface compliant with IEEE 802.3-2002 [33] is available through an RJ-45 connector and operates at 10/100 Mb/s. The direct current section of the measurement and multiplexing board supplies regulated rails at 5 V and 3.3 V for the sensing subsystems.

Excitation and measurement paths are routed through a switching matrix composed of 32 CMOS analog keys of the Vishay DG413 family, driven by 2 Texas Instruments CD74HC154M 16-channel decoders (Dallas, TX, USA). The matrix is arranged as an H bridge so that all non-active lines are referenced to ground. Excitation paths and measurement paths are switched by separate keys in order to limit crosstalk and leakage. Power for the control electronics is provided at 5 V, while peripheral circuits are supplied from the 5 V and 3.3 V rails as required.

The tomograph device was connected to the vessel outfitted with the 32-electrode array. Two rings of sixteen electrodes provided 32 channels for current injection and potential measurement. A single 16-electrode ring configuration was used, despite the availability of two rings, for several key reasons. Primarily, this choice enforces a two-dimensional (2D) cross-sectional assumption, which was essential for maintaining a fixed and repeatable boundary for training the ResNet50 reconstructor. This approach also avoids confounding vertical effects that would otherwise necessitate a more complex 3D or multi-ring forward model and complete network retraining.

Operationally, the single-ring configuration offered significant practical advantages, including simplified calibration, reduced data acquisition and processing complexity, and improved run-to-run comparability under controlled stirring. We acknowledge that this configuration results in a lower angular sampling density compared to a 32-electrode setup, which inherently limits fine-scale spatial discrimination. However, this trade-off was deemed acceptable as the primary objective was the robust recovery of machine-readable state variables (e.g., average substrate concentration) rather than high-resolution imaging. For this application, which relies on spatially averaged data, the 16-electrode ring provided the necessary stability and accuracy. We have clarified this rationale and its impact on spatial resolution in the manuscript and note that extending the system to dual-ring or full 3D acquisition is a planned direction for future work.

The instrument operated at a frequency of one kilohertz and applied an excitation current of two thousand microamperes to each driven electrode. The fixed geometry of the electrode rings and the repeatable stirring conditions ensured comparability across runs and supported time-resolved analysis of conductivity changes within the liquid. Reconstruction of conductivity fields from EIT measurements is performed with a convolutional ResNet50. Features extracted from the reconstructed images are subsequently supplied to the kinetic model described later (Section 2.3), which is used for state estimation and short-term prediction of the fermentation process.

To enhance clarity, a methodological workflow diagram (Figure 3) summarizes the proposed pipeline, illustrating the sequential transformation from raw EIT measurements through neural network-based reconstruction and kinetic modeling to closed-loop supervisory control.

The choice of ResNet50 was motivated by its stable reconstruction performance across a wide range of geometries and bulk conductivities, yielding reliable estimates of key states (e.g., substrate concentration). The neural network used had an input vector of 96 voltage measurements and an output vector of 2502 values correlated with the conductivity of the finite elements of the finite element mesh (FEM): $(1 \times 96) \longrightarrow \text{ResNET50} \longrightarrow (1 \times 2502)$. The detailed method for obtaining the EIT simulation measurements necessary to train the neural network is explained in the publication [7]. The deep neural network model comprised 175 layers interconnected through 190 connections and included 214 sets of learnable parameters as well as 106 stateful parameters. The network accepted input through a designated input port and produced outputs via a final fully connected layer mapping to 2502 outputs. Training was conducted using the Adam optimizer over a maximum of 500 epochs, with mini-batches of size 64. In each epoch, the training dataset was shuffled, and variable-length sequences were accommodated by grouping inputs into batches of the maximal sequence length encountered. Validation was performed on a separate held-out dataset at intervals of every 100 iterations, and early stopping was applied if validation performance did not improve over 200 consecutive checks. The root-mean-square error (RMSE) metric was used to monitor performance, and training progress was visualized dynamically. Computations were executed on the available hardware platform, adapting to Graphics Processing Unit (GPU) resources. During training, gradients were backpropagated through the entire network (including recurrence or temporal dependencies where present), and parameter updates followed the Adam update rules. The early stopping mechanism prevented overfitting by halting training when further refinements on the validation set stalled. All layers, residual connections, and internal transformations (e.g., convolutional, normalization, and activation) were optimized simultaneously under this regimen. Figure 4 shows the training performance of the ResNET50 through RMSE. Convergence of the error lines for the training and validation sets indicates the absence of overfitting. The green triangle indicates the iteration for which the model generated the smallest RMSE error. Training was automatically stopped by a specified early stopping condition, which included 200 consecutive epochs without decreasing the loss function value. The regular, hyperbolic shape of the graph and the lack of fluctuations indicate the absence of overfitting.

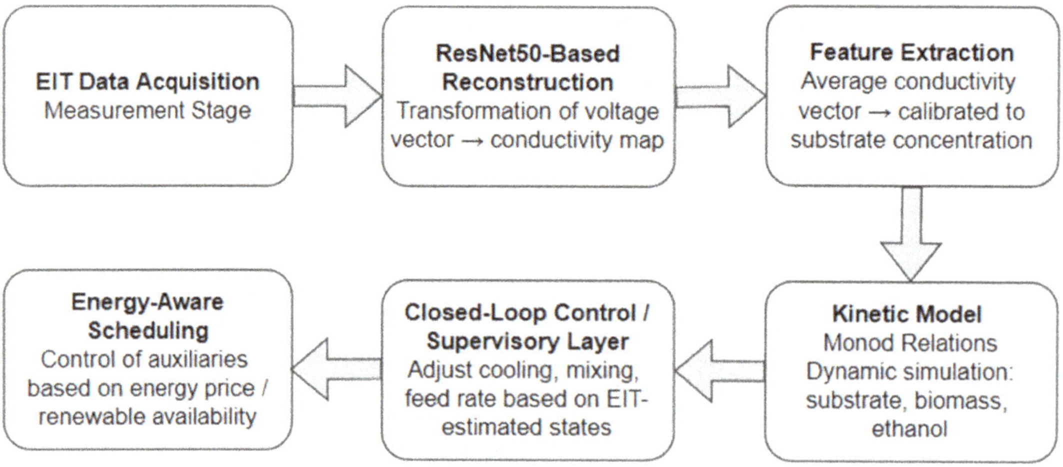

Figure 3. Methodological workflow of the proposed EIT-based monitoring and control framework.

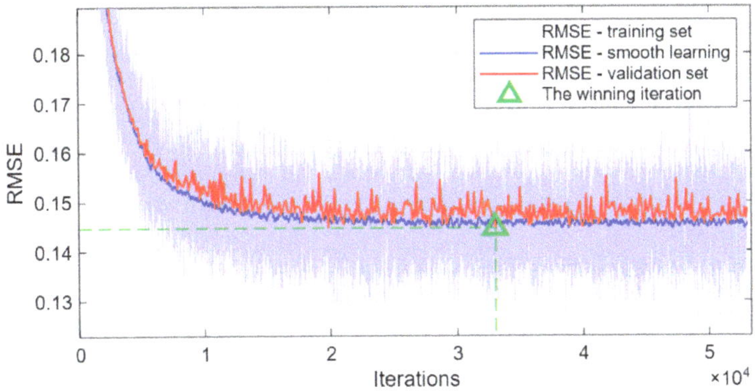

Figure 4. Training performance of ResNET50.

2.3. Simulation Environment

The simulation model shown in Figure 5 implements a dynamic closed feedback loop in which the course of alcoholic fermentation is simulated using information about the spatial distribution of electrical conductivity inside the reactor. The workflow begins by loading a conductivity map from a database of tomographic reconstructions. The map represents local conductivity values on a finite element mesh of 2502 cells/pixels and serves as an indirect indicator of the distribution of dissolved species in the broth. The *getSigma* block retrieves the current conductivity vector from the base workspace and provides it to the Simulink model. It ensures a fixed output size and allows the simulation to use the most recent conductivity data updated by other blocks, such as *trigger_plot_sfun*. The *conductivityToConcentration* block converts the conductivity distribution into a state vector describing the process conditions. It applies a linear calibration to estimate substrate concentration, then sums the results to obtain its overall value in the reactor. Biomass and ethanol concentrations are set to constant reference values. The resulting vector is used as input for *Fb_function* and the Integrator, linking tomographic data with the dynamic fermentation model and enabling EIT-based simulation. The *trigger_plot_sfun* block monitors the process state and triggers updates of the conductivity distribution used for EIT reconstruction. At the start of the simulation, it generates an initial conductivity field and activates the reset signal. During simulation, it checks the substrate concentration and, when it falls below a threshold, loads a new conductivity vector, updates the FEM image, and refreshes the visualization window. A cooldown mechanism prevents repeated triggering until the concentration rises again, ensuring controlled and synchronized updates of the model state. The Scope block in Simulink is used to visualize signals in real time during simulation.

The conductivity field is converted to an average substrate concentration through a linear calibration fitted to prior experiments. The calibration uses a proportionality coefficient of 0.99 kg/m^3 per unit of conductivity and an offset of -0.85 kg/m^3. In the baseline configuration, the concentrations of biomass and ethanol are fixed at 105 kg/m^3 and 0 kg/m^3. These constants stabilize the numerical scheme and isolate the dynamics of the substrate. They can be released in later variants to represent their time evolution.

The three-component concentration vector obtained from the calibrated conductivity is forwarded to a kinetic core based on Monod relations [34]. The Monod dependence used in the kinetic core is (2)

$$\mu = \mu_{max} \frac{S}{K_S + S}, \tag{2}$$

where $\mu_{max} = 0.0167\ \mathrm{h}^{-1}$ is the maximal specific growth rate scaled to a 46 h simulation window and $K_S = 100\ \mathrm{kg/m^3}$ is the substrate saturation constant. Stoichiometric yields that link substrate consumption to biomass formation and ethanol production are 0.1 for biomass on substrate and 0.5 for ethanol on substrate [34]. The model computes instantaneous rates from these parameters and advances the state in time by numerical integration. The Integrator block performs continuous-time integration of the input signal, updating the process state over simulation time. It is configured with an external reset that reinitializes the state whenever a rising edge signal is detected, allowing the system to restart from new initial conditions when required.

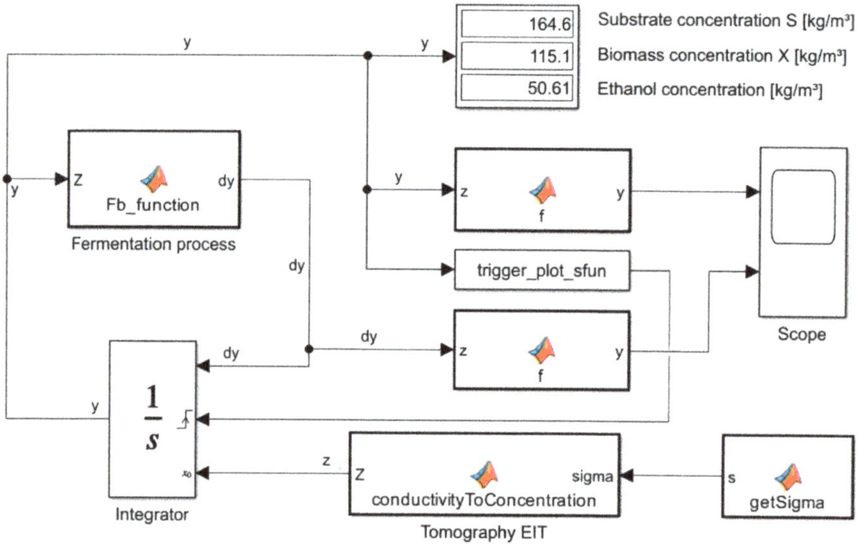

Figure 5. Schematic diagram of the simulation model of the alcoholic fermentation process.

Based on the computed specific growth rate, the reaction rates are obtained as biomass formation $r_X = \mu \cdot X$, substrate consumption $r_S = {}^{r_X}/_{Y_{XS}}$, and ethanol formation $r_P = r_S \cdot Y_{PS}$. The output of the kinetic block is the three-dimensional vector of time derivatives corresponding to changes in concentrations: substrate loss $(\mathrm{d}S/\mathrm{d}t = -r_S)$, biomass growth $(\mathrm{d}X/\mathrm{d}t = r_X)$ and ethanol production $(\mathrm{d}P/\mathrm{d}t = r_P)$. This vector is then passed to the integrator block in the Simulink model, which allows for a continuous evolution of the system state over time.

An event mechanism introduces realistic variability in internal conditions. If the simulated substrate concentration falls below $10\ \mathrm{kg/m^3}$ the model draws a new conductivity map from the reconstruction database and inserts it into the next simulation step. This refresh emulates the evolving physical and chemical environment inside the fermenter and prevents the trajectory from locking into a single spatial pattern. The update is protected by cooldown logic so that two updates cannot occur in direct succession. This avoids excessive perturbations and maintains numerical stability.

The MATLAB Simulink R2025a interface is configured to stream conductivity maps during execution. A block retrieves the current map from the MATLAB base workspace and enforces a fixed vector length equal to the mesh size. A companion block converts the conductivity to concentrations using the specified linear calibration and outputs a three-component state vector. A supervisor block monitors the current state and issues a trigger when the threshold condition is met. The visualization routine reuses an existing plot window and redraws the finite element field each time a new map is injected. This provides a continuous view of the evolving distribution throughout the run.

3. Results

3.1. Simulation Results for the Fermentation Model

Figure 6 shows the results of a dynamic simulation of alcoholic fermentation developed in Simulink and driven by electrical impedance tomography reconstructions. The model reproduces a repeated batch fermentation cycle regime in which conductivity maps inform state updates and a threshold rule initiates a new cycle. The top panel shows the substrate concentration S and ethanol concentration P versus time. The red curve documents a stepwise decrease in S during each cycle. When S reaches the threshold of about 10 kg/m^3, the supervisor block refreshes the conductivity field and the effective substrate level is reset, starting the next cycle. The blue curve shows the corresponding increase in P within a cycle, followed by a reset at the cycle boundary, which represents the start of a new controlled batch fermentation cycle. The lower panel shows the time derivatives dS/dt and dP/dt. The red derivative is negative and becomes more pronounced during active consumption and then returns to zero after the refresh event. The blue derivative is positive during production and follows the expected cycle profile. The periodic patterns confirm that the threshold-triggered conductivity map update produces stable and repeatable trajectories consistent with the kinetic assumptions and control logic.

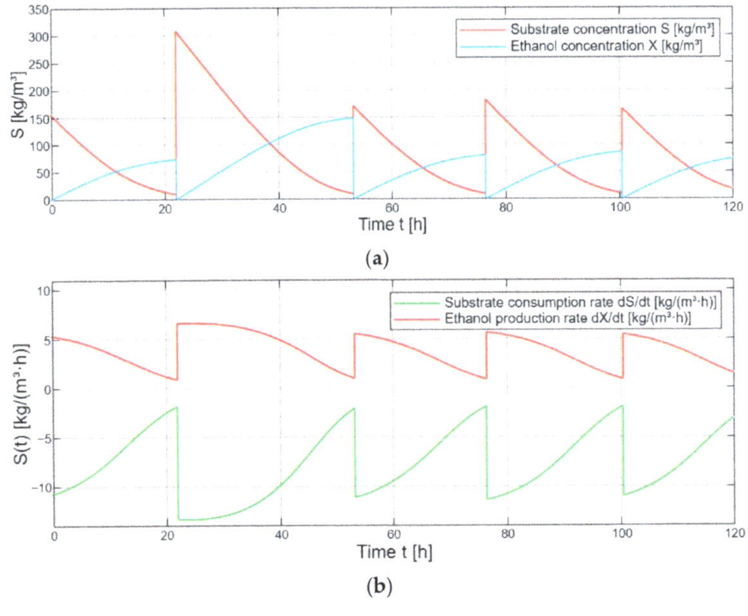

Figure 6. Simulation graph of substrate and ethanol during the multiple fermentation process: (**a**) substrate (S) and ethanol (X) concentrations vs. time [kg/m^3]; (**b**) through substrate consumption/production rates.

Figure 7 shows a single uninterrupted batch fermentation cycle over 46 h with an initial substrate concentration of $S_0 = 285.7$ kg/m^3 and Monod kinetics as specified in the simulation environment. The top panel shows $S(t)$ and $P(t)$. The red curve declines monotonically, with the rate of decline slowing as the substrate is depleted. The blue curve rises towards an asymptote as the conversion proceeds and approaches a plateau when S is nearly exhausted. This behavior is consistent with the expected course of a batch fermentation without replenishment. The lower panel shows dS/dt and dP/dt. The magnitude of dS/dt is greatest at the beginning, when the driving concentration is high, and then decreases as the process nears completion. The production rate dP/dt increases to a peak during the most active metabolic phase and then decreases as the substrate becomes limiting.

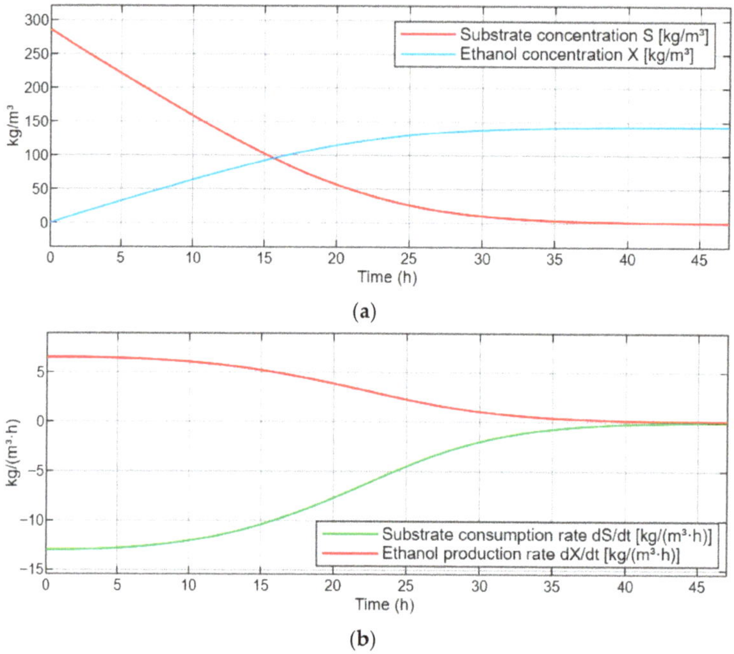

Figure 7. Simulation graph of a 46 h alcoholic fermentation process with an initial substrate concentration of $S_0 = 285.7$ kg/m^3; (**a**) substrate (S) and ethanol (X) concentrations vs. time [kg/m^3]; (**b**) through substrate consumption/production rates.

Taken together, the results in Figures 6 and 7 validate the simulation workflow. The repeated batch scenario demonstrates robust interaction between imaging-informed state updates and process kinetics. The single batch scenario confirms that the parameterization reproduces classical batch behavior under Monod dependence. Both cases show smooth evolution of concentrations and derivatives, predictable responses to the threshold rule, and consistency with mass balance expectations, which supports the use of the model for analysis and prospective decision support in controlled fermentation studies.

3.2. Validation of the Alcohol Fermentation Monitoring Model

The validation pathway begins with a simplified simulation of the real fermentation course implemented in Simulink and summarized in Figure 8. This configuration reproduces the time evolution of substrate and ethanol by combining a matrix of precomputed electrical impedance tomography reconstructions with a predefined time law governing the evolution of process parameters. In contrast to the dynamic Monod-based simulator presented earlier, which advances the state by integrating differential equations, the validation model prescribes concentrations directly as functions of time and uses the images as a consistent background for visualization and parameter readout. The MATLAB Function block marked as *updateReconstruction* receives simulation time and the reconstruction matrix and returns the current pair of concentrations for substrate and ethanol. The *sigma_table* variable provided from the MATLAB base workspace contains EIT reconstructions performed at specific time intervals during the 46 h fermentation process under study.

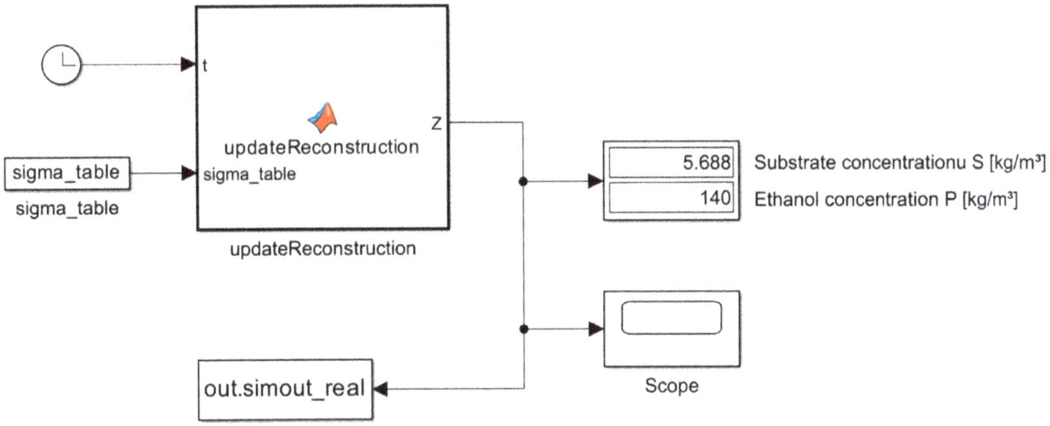

Figure 8. Schematic Simulink model of the actual course of the alcoholic fermentation process.

Quantitative validation proceeds in two steps. First, an analytical reference trajectory is generated by an exponential law for substrate decay and by a mass balance for ethanol formation as implemented in the update block. The substrate follows (3)

$$S(t) = S_0 \cdot \exp\left(-\frac{t}{\tau}\right), \tag{3}$$

with $S_0 = 285.7\ \mathrm{kg/m^3}$ and $\tau = 12\,\mathrm{h}$ over a window of approximately forty-six hours. The ethanol is computed from the mass balance (4)

$$P(t) = Y \cdot (S_0 - S(t)), \tag{4}$$

with yield $Y = 0.\,5$. These expressions define the reference pair of time series used for comparison and for driving the simplified simulator shown in Figure 8.

Second, the analytical profiles are confronted with concentrations recovered from real electrical impedance tomography measurements. A direct comparison between analytical simulation and experimental recovery is presented in Figure 9. The dashed red and blue curves show the simulated reference, and the red and blue markers show the values inferred from electrical impedance tomography. The agreement is close in both shape and scale. Early time differences are small and can be explained by mixing transients and local heterogeneity in conductivity. Late time differences are also small and are consistent with the finite spatial resolution of the reconstruction and with the decreasing signal-to-noise ratio as conductivity gradients shrink. The overall match supports the adopted conductivity to concentration calibration and confirms that the simplified simulator is an adequate reference for validating the monitoring pipeline.

It should be emphasized that the substrate and ethanol concentration values presented in Figure 9 were not measured using reference methods. The term "actual measurements" refers to EIT tomographic electrical conductivity quantifications performed during a 46 h alcoholic fermentation process. Each measurement point on the graph represents the conversion of a 2502-element tomographic reconstruction vector (sigma) to the corresponding chemical concentrations using a calibration function. The regularity of the point pattern results from the nature of the conversion method used and the spatial averaging inherent in tomographic measurements. EIT measures averaged conductivity across the entire reactor volume, which naturally reduces local fluctuations and measurement noise typical of point-based analytical methods. It should be emphasized that the measurement points presented in Figure 9 represent averaged values across the entire reactor volume, which significantly reduces the influence of local fluctuations and measurement noise.

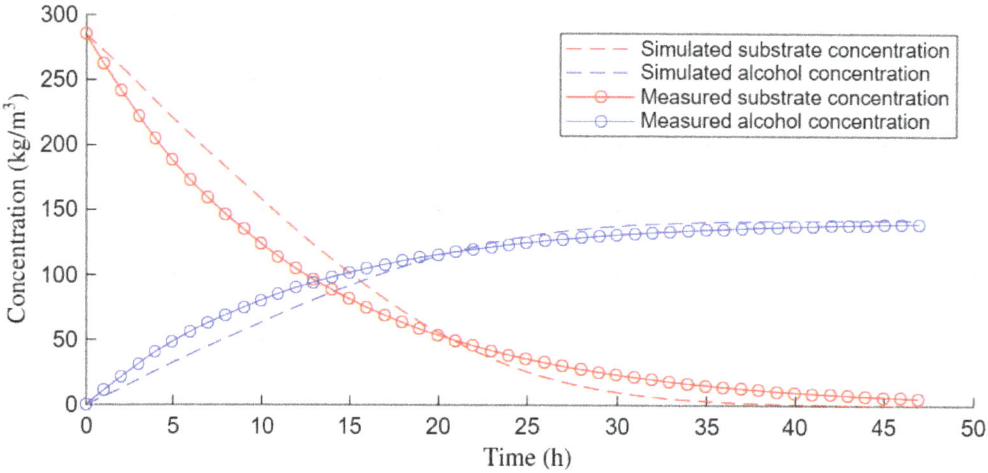

Figure 9. Comparison of the 46 h alcoholic fermentation process between simulations and actual EIT measurements.

Figure 10 shows a sequence of 12 tomographic reconstructions of the tested reactor taken during a 46 h fermentation process. Reconstruction (a) was taken at the beginning of the process, while reconstruction (l) was taken at its completion. The sequence reveals a progressive redistribution of conductivity consistent with sugar depletion and ethanol formation. Color images adapted to human perception are for informational purposes only. The numerical values of the output vector are important for the process control system. Based on these values, the algorithm can estimate substrate concentrations. The varied colors confirm that the imaging layer provides stable and interpretable inputs to the concentration estimation step.

We evaluated computational performance on an Asus TUF laptop with an AMD Ryzen 7 5800H CPU at 3.2 GHz, 16 GB RAM, and an NVIDIA GeForce RTX 3060 with 6.4 GB VRAM running Windows 11. For the 16-electrode protocol at 1 kHz, one full EIT frame required 28–34 ms with GPU inference and 90–110 ms with CPU-only execution. The control layer added 20–30 ms from feature extraction through Monod model update to command dispatch. The end-to-end latency measured from the last voltage sample to actuation was 55–70 ms with the GPU enabled and 130–150 ms on CPU only, providing stable real-time operation at approximately 10–15 Hz with the GPU and 5–7 Hz on CPU. In practice, the system runs on any recent 64-bit machine with at least a quad-core CPU and 8 GB RAM, while a CUDA-capable GPU with 4 GB or more VRAM is recommended to reach the higher update rates.

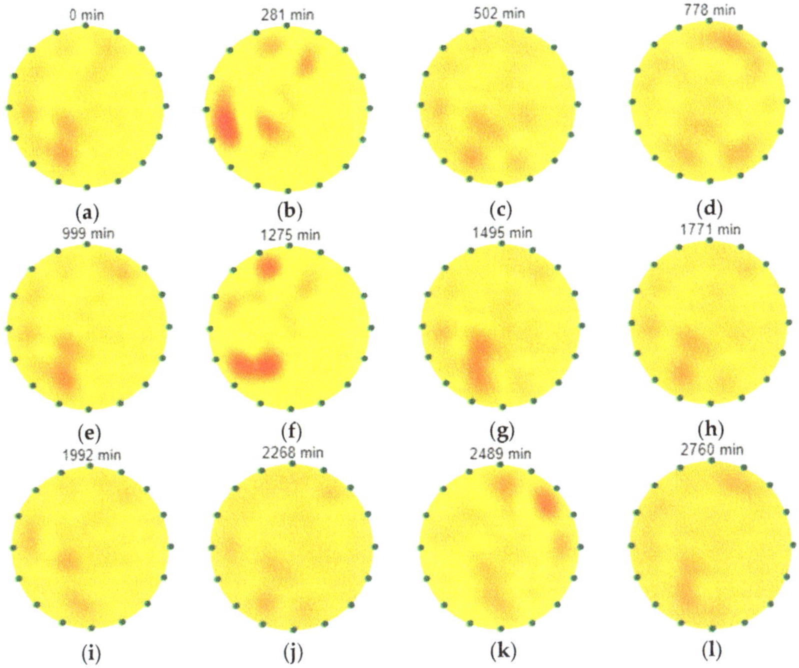

Figure 10. Sequence of tomographic cross-sections acquired during a 46 h fermentation. Reconstructions (**a–l**) show successive time points in minutes from the process start (t = 0).

3.3. Energy Accounting for EIT-Supervised Fermentation: Models and Results

Below, the energetic implications of coupling EIT-based state estimation with phase-aware temperature targeting are quantified. Building on the previously introduced kinetic core and conductivity-to-concentration calibration, a control-energy surrogate is derived that links setpoint trajectories to compressor and agitation loads via a temperature-lift–dependent COP, passive (free) cooling, and a cube-law model of mixing. An EIT-supervised strategy—whose targets are confined to admissible, phase-dependent temperature bands and biased toward ambient—is then compared with three traditional programs (constant, linear, exponential) over a 46 h batch. The resulting time profiles of instantaneous electric power, target temperatures with bands, and metabolic heat release are integrated to obtain cumulative energy, thereby allowing the specific contributions of supervision, free cooling, and reduced mixing to the observed savings to be isolated.

Figure 11 aggregates the dynamic response of four temperature control strategies over a 46 h alcoholic fermentation and the underlying metabolic drive, using the same color code across panels (EIT-supervised—blue, constant 25 °C—red, linear—green, exponential—magenta). In the top panel ("Instantaneous Electric Power"), each curve represents the total electric load $P_{\mathrm{tot}}(t) = P_{\mathrm{cool}}(t) + P_{\mathrm{mix}}(t)$, where $P_{\mathrm{cool}}(t)$ compressor (cooling) electric power [W] and $P_{\mathrm{mix}}(t)$ agitator/mixer electric power [W].

For the EIT-supervised case (blue), cooling is dispatched only when tomographic state estimation indicates that temperature would leave a phase-dependent admissible band, and mixing setpoints are reduced outside the peak metabolic phase. Consequently, the blue power trace exhibits short plateaus and long low-duty intervals. In the three traditional setpoint programs, agitation is kept constant while the cooling duty decays mainly because the commanded temperature gradually approaches the ambient. The 25 °C hold declines the slowest (red), the linear ramp declines linearly (green), and the exponential target produces a concave decay (magenta). Cooling power in all cases is computed with a temperature-lift-dependent coefficient of performance (COP) and a "free-cooling" term that removes heat to ambient without compressor work when permissible, i.e.,

$$COP(T_b, T_c) = \max\left(COP_{min}, COP_0 - k_{COP} \cdot [T_b - T_c]_+\right) \quad (5)$$

$$Q_{pass} = UA \cdot [T_b - T_{amb}]_+ \quad (6)$$

$$Q_{need} = \max(0, U\,A \cdot (T_b - T_c) - \alpha \cdot Q_{met}(t)) \quad (7)$$

$$P_{cool}(t) = \frac{\max(0, Q_{need} - Q_{pass})}{COP(T_b, T_c)} \quad (8)$$

where: $COP(T_b, T_c)$—coefficient of performance of the refrigeration cycle [–], depends on the temperature lift, COP_{min}—lower bound of the COP [–], COP_0—nominal COP intercept (for negligible lift) [–], k_{COP}—COP degradation coefficient per kelvin of lift [K^{-1}], T_b—broth (process) temperature [°C or K], T_c—coolant or jacket-side temperature used in the lift [°C or K] (often cooling-water supply), Q_{pass}—passive ("free-cooling") heat rejected to the ambient without compressor work [W], UA—overall heat-transfer conductance (U × A) [$W \cdot K^{-1}$], T_{amb}—ambient air temperature [°C or K], Q_{need}—gross cooling duty before accounting for passive removal [W], α—fractional coefficient (0–1) crediting metabolic heat against external cooling [–], $Q_{met}(t)$—metabolic heat-release rate [W].

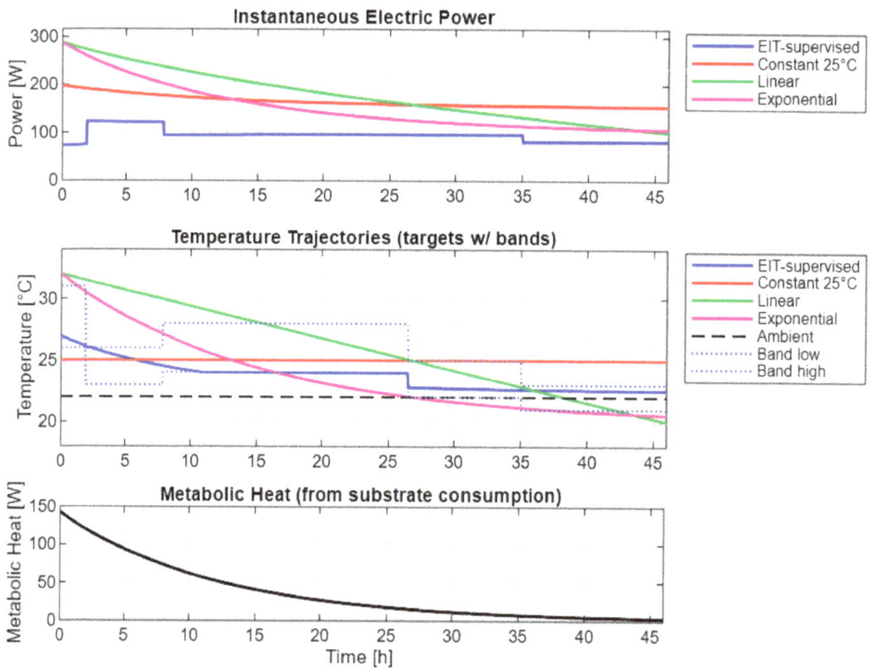

Figure 11. Instantaneous power, temperature targets, and metabolic heat under EIT-supervised vs. traditional control.

The mixing power follows a variable-frequency-drive cube law, $P_{mix}(t) = P_{base}\cdot(n(t)/n_0)^3$, with $n(t)$ lowered by the supervisor during lag and maturation. These relationships are part of the control-energy surrogate model and complement the kinetic core presented in the manuscript.

The middle panel in Figure 11, titled "Temperature Trajectories (targets w/ bands)", shows the target temperature programs that feed the above power calculation. The three traditional strategies (red, green, magenta) are predefined setpoints. By contrast, the EIT-supervised trajectory (solid blue) is constrained to remain between two dotted blue envelopes labeled "Band low" and "Band high". In the code, these are *Tband_lo* and *Tband_hi*. They are phase-dependent admissible corridors inferred from the tomographically estimated state: EIT reconstructions are converted to concentrations via a calibrated map and propagated through a Monod kinetic core, which provides a smooth phase indicator (lag → early exponential → peak → late → maturation). The conductivity-to-concentration calibration and the Monod dependence used by the kinetic block, $\mu = \mu_{max}\cdot S/(K_S + S)$, with $\mu_{max} = 0.0167\ h^{-1}$ and $K_S = 100\ kg/m^3$, together with the associated rate equations $r_X = \mu X$, $r_S = r_X/Y_{XS}$, $r_P = r_S\cdot Y_{PS}$ and balances $\dot{S} = -r_S$, $\dot{X} = r_X$, $\dot{P} = r_P$ (Sections 2 and 3), are employed. The analytical reference trajectory used in validation, $S(t) = S_0\cdot e^{-t/\tau}$ with $S_0 = 285.7\ kg\cdot m^{-3}$ and $\tau = 12$ h, and $P(t) = Y\cdot(S_0 - S(t))$ (Equations (2)–(4) in the manuscript), is also provided. These elements underpin our phase logic and, therefore, the blue bands and the supervised temperature trace.

The bottom panel ("Metabolic Heat (from substrate consumption)") depicts the driving heat release $Q_{met}(t)$, which declines monotonically as substrate is depleted. In the kinetic core documented in the manuscript, rates follow Monod and stoichiometric yields. For the energy accounting used here, we map the consumption rate to heat release via the surrogate $Q_{met}(t) = k_Q\cdot\max\left(0, -\dot{S}(t)\right)$, which is consistent with the simulated behavior of $\dot{S}$ and $\dot{P}$ reported in the validation figures (the magnitude of $\dot{S}$ is the largest early and decreases as the batch nears completion, while $\dot{P}$ peaks at mid-run).

Figure 12 summarizes the integral energetic outcome. Each bar reports the 46 h electric energy $E = \int_0^{46h} P_{tot}(t)dt$.

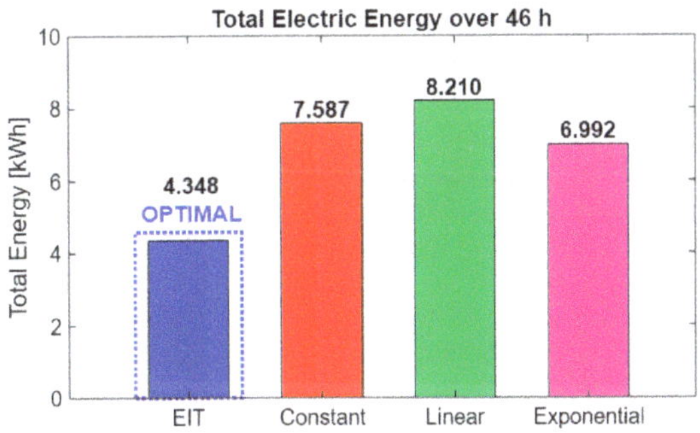

Figure 12. Energy use comparison: EIT-supervised vs. traditional setpoints (46 h).

The EIT-supervised strategy (blue) attains the minimum (~4.35 kWh) because the supervisor biases the target toward ambient within the admissible corridor, which reduces temperature lift and increases COP, exploits free-cooling whenever $T_b > T_{amb}$ to offload compressor work, and lowers agitation outside the peak metabolic window according to the cube law. The constant 25 °C hold (red, ~7.59 kWh) is penalized by sustained lift. The linear ramp (green, ~8.21 kWh) incurs high early-time duty when the setpoint is far from ambient, and the exponential program (magenta, ~6.99 kWh) moderates duty relative to linear but remains less efficient than EIT supervision because it ignores the real-time metabolic state reflected by EIT. Altogether, Figures 11 and 12 illustrate how the EIT-driven pipeline (conductivity field → calibrated concentrations → Monod kinetics → phase-aware temperature bands) translates into lower instantaneous power and the best cumulative energy over the batch.

3.4. Sensitivity and Uncertainty Analysis of the Conductivity—Substrate Calibration Model

Figure 13 presents the sensitivity and uncertainty analysis of the linear calibration model that relates electrical conductivity σ (in arbitrary units [a.u.]) to substrate concentration S during fermentation. Panel (A) shows the calibration curve obtained from experimental data, where measured conductivity values are plotted against analytically determined substrate concentrations. The solid black line represents the fitted linear model described by the equation $\mathrm{S} = 0.986 \cdot \sigma - 1.671\ \left[\mathrm{kg{\cdot}m^{-3}}\right]$, while the dashed and dotted lines indicate the 95% confidence interval (CI) and 95% prediction interval (PI), respectively. The color of the data points corresponds to the process temperature (T), highlighting its relatively minor variation during the experiment. The linearity of the relationship confirms that conductivity can be used as a reliable proxy for substrate concentration within the tested range.

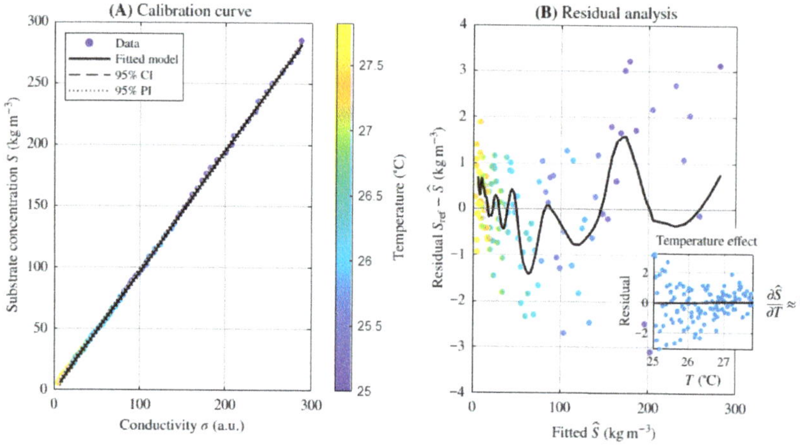

Figure 13. Sensitivity and uncertainty analysis of the linear calibration between conductivity and substrate concentration: (**A**)—calibration curve for $S = 0.986{\cdot}\sigma - 1.671\ \left[\mathrm{kgm^{-3}}\right]$, (**B**)—residual analysis with temperature-related bias visualization.

Panel (B) provides a residual analysis to assess model robustness and to visualize temperature-related effects. The residuals, defined as $S_{\mathrm{ref}} - \hat{S}$, are plotted against the fitted values, where S_{ref} denotes the reference substrate concentration and $\hat{S}$ the model estimate. The black curve represents a locally weighted LOESS regression, showing small systematic deviations that suggest minor nonlinearity at the extremes of the range. The inset plot, labeled "Temperature effect," depicts the residuals as a function of temperature T [°C] and shows a weak positive slope, approximately $\partial\hat{S}/\partial T \approx 0.037$ [kg m^{-3} °C^{-1}], indicating that conductivity slightly increases with temperature. Together, these panels demonstrate that the calibration model provides accurate concentration estimates with limited temperature sensitivity and acceptable uncertainty for process monitoring applications.

Figure 14 summarizes the Monte Carlo uncertainty decomposition for the linear conductivity–substrate calibration at a representative operating point of $S \approx 150$ kg m^{-3}.

The bar chart reports the standard uncertainty contributed by temperature variation, measurement noise, and electrode drift, together with the combined effect. Temperature dominates with an estimated 3.00 kg m^{-3} one-sigma contribution, followed by measurement noise at 1.50 kg m^{-3} and electrode drift at 0.76 kg m^{-3}, yielding a combined standard uncertainty of 3.45 kg m^{-3}. These values were obtained using a physically plausible disturbance model that assumes approximately 1 °C one-sigma temperature variability with a 2% per °C conductivity coefficient, 1% measurement noise, and 0.5% slow electrode drift, which reflects the operating window of our experiments. Expressed relatively, the combined uncertainty corresponds to about 2.3% at $S = 150$ kg m^{-3} and about 1.2% of the initial concentration $S_0 = 285.7$ kg m^{-3}, indicating that the calibration maintains sufficient precision for closed-loop supervision and event-triggered control. The figure complements the calibration and residual analyses by quantifying how each disturbance source propagates to concentration estimates and by confirming that the overall uncertainty remains small compared with typical decision thresholds used in our control logic.

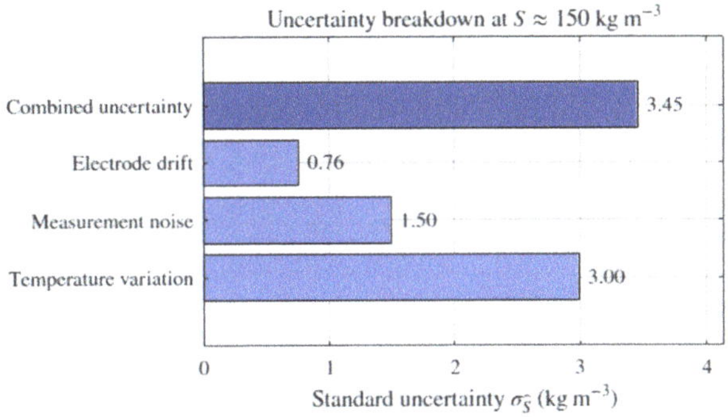

Figure 14. Uncertainty breakdown of the conductivity–substrate calibration.

4. Discussion

The study demonstrates an AI-enhanced monitoring pipeline for alcoholic fermentation that couples electrical impedance tomography with a residual convolutional network from the ResNet family and a Monod-based kinetic core to estimate substrate and ethanol and to enable short-range prediction. The approach is consistent with the recent progress in deep learning for electrical impedance tomography reconstruction and with reviews that document the ill-posed nature of the inverse problem and the value of data-driven priors [24,35,36].

The sensing layer builds on the use of impedance-based measurements in bioprocess monitoring, where noninvasive electrical methods track biomass and composition during fermentation. Evidence for feasibility is provided by demonstrations of electrochemical impedance spectroscopy in microbial cultivations and by reports that impedance spectroscopy can follow biomass in complex broths [37–39].

Validation uses analytical reference trajectories grounded in standard Monod-style modeling of ethanol fermentation and a mass balance for the product. The observed agreement between analytical references and concentrations recovered from electrical impedance tomography aligns with recent kinetic studies of ethanol production that employ unstructured models in batch and fed-batch regimes. Early deviations can be ascribed to mixing transients and spatial heterogeneity, and late deviations are compatible with the known resolution and signal-to-noise limits of electrical impedance tomography [40,41].

The batch course reported here is consistent with the established physiology of *Saccharomyces cerevisiae* where rapid early sugar uptake is followed by ethanol accumulation and rate deceleration under substrate depletion and product stress. The broader literature documents cellular responses that underpin ethanol tolerance and membrane level adaptations during ethanol challenge and these mechanisms explain the observed turning points of the trajectories [42,43].

The sensitivity of fermentation kinetics to nitrogen availability is well described by controlled studies and reviews that resolve effects of both level and timing of yeast assimilable nitrogen. The stable mapping from impedance signals to concentration estimates observed here is therefore coherent with scenarios where nitrogen regimes modulate rates and product formation while the electrical pathway offers a continuous surrogate for state estimation [44–46].

The methodological choice to pair electrical impedance tomography with a residual convolutional reconstructor follows current practice in imaging-guided state estimation and is supported by surveys that compare single-network reconstruction and hybrid strategies for this modality. The stability of the repeated batch regime with an event-triggered refresh is consistent with the view that imaging-informed signals can support supervisory control and scheduling in bioethanol processes within the accuracy bounds of modern electrical impedance tomography [24,36].

The proposed AI-enhanced sensing and estimation enables supervisory scheduling of electricity use in fermentation and is consistent with evidence that process industries can supply demand-side flexibility to support grids with variable renewable generation [47]. Brewery and fermentation sites carry significant electric loads in refrigeration and agitation, and refrigeration is often the dominant electrical consumer that can be time shifted without compromising product quality when process constraints are respected. The event-triggered threshold and cooldown logic provide actionable signals to shift cooling and pumping toward low price or high renewable periods, while machine learning based bioprocess monitoring supports real-time decision making [48]. When co-located with biogas units, additional flexibility can be provided because modern biogas plants can modulate output in response to market conditions and renewable variability [49]. Digitization of these functions should follow established guidance for operational technology security to maintain resilience of scheduling and control [50].

Taken together, the results indicate that an imaging-informed AI pipeline can deliver reliable state estimation and short-range prediction for alcoholic fermentation under realistic variability, with observed deviations consistent with known limits of electrical impedance tomography and process mixing. The approach enables energy-aware supervision in which cooling and agitation are scheduled against price signals and renewable availability, and it can be extended where fermentation is co-located with bioenergy units.

Measurement noise, electrode drift, and process disturbances are inherent challenges in EIT-based control systems and directly influence reconstruction stability and control reliability. Measurement noise primarily affects the voltage acquisition stage, introducing random fluctuations that propagate through the inverse problem and can cause small but noticeable variations in reconstructed conductivity maps. These effects are mitigated through the use of averaging filters, regularization, and deep learning based reconstruction models, such as the ResNet50 applied in this work, which are inherently robust to high-frequency noise. Electrode drift and polarization introduce low-frequency systematic errors, particularly during long-term experiments, leading to gradual baseline shifts in reconstructed conductivity. This issue is alleviated through periodic impedance calibration, temperature compensation, and the use of non-polarizable electrodes. Process disturbances such as mixing inhomogeneities, gas bubble formation, or thermal gradients can transiently distort the electric field distribution and thus perturb the reconstructed conductivity. The control algorithm mitigates these effects through temporal smoothing of feature vectors and event-triggered decision logic, ensuring that control actions are based on stable trends rather than instantaneous fluctuations. Overall, the combined use of noise-tolerant reconstruction, adaptive calibration, and disturbance-aware control logic provides sufficient stability for reliable closed-loop operation under realistic industrial conditions.

Beyond the controlled laboratory conditions, the framework is designed to remain robust under typical industrial disturbances by combining sensing-layer compensation with disturbance-aware decision logic. Sudden temperature excursions are handled through continuous multi-point temperature monitoring and compensation in the conductivity→concentration mapping, with supervisory guards that delay actuation and trigger rapid re-baselining if deviations exceed preset limits. Electrode fouling and polarization are managed by periodic impedance checks and baseline tracking. The estimator updates gain/offset terms online and can switch to a diagnostic mode that suspends control actions when quality indicators (e.g., reconstruction consistency and residual thresholds) degrade, which addresses confounders known for EIT in industrial media. At the control layer, temporal smoothing, hysteresis, and a cooldown mechanism ensure that state estimates must remain stable over a dwell time before any actuation is issued, preventing spurious responses to transient artefacts caused by bubbles, mixing inhomogeneities, or drift. Together, these measures provide a fail-safe operating envelope in which the system either adapts to disturbances or reverts to safe monitoring until recalibration or maintenance restores nominal conditions.

Key limitations of the model include simplified kinetic assumptions, the scale and geometry of the experimental rig, and the specificity of the calibration. In the baseline configuration, the Monod-type kinetic core adopts simplified assumptions—biomass and ethanol concentrations are held fixed at 105 $kg \cdot m^{-3}$ and 0 $kg \cdot m^{-3}$, respectively—which reduces model fidelity with respect to coupled metabolic dynamics, product inhibition and nutrient-driven physiological changes that commonly occur in industrial fermentations. The experimental validation was performed in a bench-scale 3.5 L tank reactor with a fixed electrode geometry and repeatable stirring. Such a small, well-mixed tank does not fully reproduce the mixing patterns, heat transfer characteristics, and three-dimensional flow structures of industrial-scale fermenters, and the model's assumption of homogeneous conditions after mixing therefore underestimates the spatial heterogeneity often present in production reactors. These advances would position fermentation as a flexible demand resource that supports stable and efficient operation of renewable integrated power systems.

5. Conclusions

The motivation for using Electrical Impedance Tomography (EIT) in closed-loop process control lies in its ability to transform raw boundary measurements into a spatially resolved representation of the process interior by solving the tomographic inverse problem. The reconstruction yields a pixelwise conductivity map, where each pixel is associated with a well-defined spatial location. This representation can be conveniently expressed as a numerical output vector whose component order encodes the spatial distribution of process variables. Such a vector can be further processed in a deterministic or data-driven manner—using statistical inference, fuzzy logic, or machine learning—to estimate key control parameters, including substrate concentrations, ion content (pH), or biomass levels. These quantities form the basis for control actions such as adjusting feed rates, mixing intensity, or temperature setpoints. Importantly, without the intermediate tomographic reconstruction step, direct measurement signals alone would provide only aggregated or localized information and would be insufficient to achieve comparable spatial resolution and estimation accuracy. EIT thus enables the transition from conventional point-based monitoring to full-field, model-based control, supporting automation and facilitating operation under Industry 4.0 and AI-enhanced paradigms.

The adoption of EIT as a core sensing modality fundamentally expands the capabilities of process automation, making it possible to achieve high-fidelity, real-time state estimation that is both robust and scalable. This paradigm supports the development of self-optimizing, unattended control systems capable of dynamically responding to process disturbances and external energy signals. As a result, EIT-based solutions offer a promising pathway toward improving process efficiency, reducing energy consumption, and enabling more sustainable and resilient industrial operations—objectives that are particularly relevant in the era of renewable energy integration and distributed energy management.

Author Contributions: Conceptualization, K.K. and T.R.; methodology, G.K., T.R. and K.K.; software, K.K. and T.R.; validation, K.G., E.G. and A.S.; resources, T.R.; data curation, T.R. and G.K.; writing—original draft preparation, M.K. and G.K.; writing—review and editing, T.R., M.K., K.G., E.G. and A.S.; visualization, K.K. and G.K.; supervision, T.R.; funding acquisition, T.R. All authors have read and agreed to the published version of the manuscript.

Funding: This research received no external funding.

Data Availability Statement: The data presented in this study are available on request from the corresponding author. The data is not made public because it is the property of the company.

Conflicts of Interest: Authors Krzysztof Król and Tomasz Rymarczyk were employed by the company Research and Development Center, Netrix S.A. The remaining authors declare that the research was conducted in the absence of any commercial or financial relationships that could be construed as a potential conflict of interest.

Abbreviations

The following abbreviations are used in this manuscript:

AI	Artificial Intelligence
a.u.	arbitrary units
ATP	adenosine 5′-triphosphate
CMOS	Complementary Metal Oxide Semiconductor
ECT	Electrical Capacitance Tomography
EIT	Electrical Impedance Tomography
FEM	Finite Element Method
GPU	Graphics Processing Unit
LCD	Liquid Crystal Display
MLOps	Machine Learning Operations
MRI	Magnetic Resonance Imaging
NIR	Near Infrared Spectroscopy
NMR	Nuclear Magnetic Resonance
OCT	Optical Coherence Tomography
PAT	Process Analytical Technology
RJ-45	Registered Jack 45
RMSE	Root Mean Square Error
UST	Ultrasound Tomography
UV-Vis	Ultraviolet Visible Spectroscopy

References

1. Su, Q.; Ganesh, S.; Moreno, M.; Bommireddy, Y.; Gonzalez, M.; Reklaitis, G.V.; Nagy, Z.K. A Perspective on Quality-by-Control (QbC) in Pharmaceutical Continuous Manufacturing. *Comput. Chem. Eng.* **2019**, *125*, 216–231. [CrossRef]
2. Gerzon, G.; Sheng, Y.; Kirkitadze, M. Process Analytical Technologies—Advances in Bioprocess Integration and Future Perspectives. *J. Pharm. Biomed. Anal.* **2022**, *207*, 114379. [CrossRef] [PubMed]
3. Bowler, A.L.; Bakalis, S.; Watson, N.J. A Review of In-Line and on-Line Measurement Techniques to Monitor Industrial Mixing Processes. *Chem. Eng. Res. Des.* **2020**, *153*, 463–495. [CrossRef]
4. Hong, C.Y.; Zhang, Y.F.; Zhang, M.X.; Leung, L.M.G.; Liu, L.Q. Application of FBG Sensors for Geotechnical Health Monitoring, a Review of Sensor Design, Implementation Methods and Packaging Techniques. *Sens. Actuators A Phys.* **2016**, *244*, 184–197. [CrossRef]
5. Roberts, J.; Power, A.; Chapman, J.; Chandra, S.; Cozzolino, D. The Use of UV-Vis Spectroscopy in Bioprocess and Fermentation Monitoring. *Fermentation* **2018**, *4*, 18. [CrossRef]
6. Kulisz, M.; Kłosowski, G.; Rymarczyk, T.; Hoła, A.; Niderla, K.; Sikora, J. The Use of the Multi-Sequential LSTM in Electrical Tomography for Masonry Wall Moisture Detection. *Measurement* **2024**, *234*, 114860. [CrossRef]
7. Kłosowski, G.; Rymarczyk, T.; Niderla, K.; Kulisz, M.; Skowron, Ł.; Soleimani, M. Using an LSTM Network to Monitor Industrial Reactors Using Electrical Capacitance and Impedance Tomography—A Hybrid Approach. *Eksploat. I Niezawodn. Maint. Reliab.* **2023**, *25*, 2023. [CrossRef]
8. Przysucha, B.; Wójcik, D.; Rymarczyk, T.; Król, K.; Kozłowski, E.; Gąsior, M. Analysis of Reconstruction Energy Efficiency in EIT and ECT 3D Tomography Based on Elastic Net. *Energies* **2023**, *16*, 1490. [CrossRef]
9. Koulountzios, P.; Rymarczyk, T.; Soleimani, M. Ultrasonic Time-of-Flight Computed Tomography for Investigation of Batch Crystallisation Processes. *Sensors* **2021**, *21*, 639. [CrossRef] [PubMed]
10. Ultrasound Tomography for Control of Batch Crystallization—The University of Bath's Research Portal. Available online: https://researchportal.bath.ac.uk/en/studentTheses/ultrasound-tomography-for-control-of-batch-crystallization (accessed on 25 August 2025).
11. Ziolkowski, M.; Gratkowski, S.; Zywica, A.R. Analytical and Numerical Models of the Magnetoacoustic Tomography with Magnetic Induction. *COMPEL—Int. J. Comput. Math. Electr. Electron. Eng.* **2018**, *37*, 538–548. [CrossRef]
12. Gocławski, J.; Sekulska-Nalewajko, J.; Korzeniewska, E. Prediction of Textile Pilling Resistance Using Optical Coherence Tomography. *Sci. Rep.* **2022**, *12*, 1–15. [CrossRef]
13. Korzeniewska, E.; Gałązka-Czarnecka, I.; Sekulska-Nalewajko, J.; Gocławski, J.; Dróżdż, T.; Kiełbasa, P. Assessment of Changes in Vitamin Content and Morphological Characteristics in Strawberries Modified with a Pulsed Electric Field Using Chromatography and Optical Coherence Tomography. *NFS J.* **2025**, *38*, 100217. [CrossRef]
14. Hlava, J.; Abouelazayem, S. Control Systems with Tomographic Sensors—A Review. *Sensors* **2022**, *22*, 2847. [CrossRef] [PubMed]
15. Hampel, U.; Babout, L.; Banasiak, R.; Schleicher, E.; Soleimani, M.; Wondrak, T.; Vauhkonen, M.; Lähivaara, T.; Tan, C.; Hoyle, B.; et al. A Review on Fast Tomographic Imaging Techniques and Their Potential Application in Industrial Process Control. *Sensors* **2022**, *22*, 2309. [CrossRef]
16. Hollenberg, M.; Liebing, T.; Kern, T.A.; Kähler, D. Simulation Framework for Electrical Impedance Tomography Systems. *Meas. Sens.* **2025**, *38*, 101416. [CrossRef]
17. Rao, G.; Aghajanian, S.; Koiranen, T.; Wajman, R.; Jackowska-Strumiłło, L. Process Monitoring of Antisolvent Based Crystallization in Low Conductivity Solutions Using Electrical Impedance Spectroscopy and 2-D Electrical Resistance Tomography. *Appl. Sci.* **2020**, *10*, 3903. [CrossRef]
18. Yan, R.; Viumdal, H.; Mylvaganam, S. Process Tomography for Model Free Adaptive Control (MFAC) via Flow Regime Identification in Multiphase Flows. *IFAC-Pap.* **2020**, *53*, 11753–11760. [CrossRef]
19. Yang, C.; Cui, Z.; Xue, Q.; Wang, H.; Zhang, D.; Geng, Y. Application of a High Speed ECT System to Online Monitoring of Pneumatic Conveying Process. *Measurement* **2014**, *48*, 29–42. [CrossRef]
20. Metcalfe, B.; Acosta-Pavas, J.C.; Robles-Rodriguez, C.E.; Georgakilas, G.K.; Dalamagas, T.; Aceves-Lara, C.A.; Daboussi, F.; Koehorst, J.J.; Corrales, D.C. Towards a Machine Learning Operations (MLOps) Soft Sensor for Real-Time Predictions in Industrial-Scale Fed-Batch Fermentation. *Comput. Chem. Eng.* **2025**, *194*, 108991. [CrossRef]
21. Wu, D.; Xu, Y.; Xu, F.; Shao, M.; Huang, M. Machine Learning Algorithms for In-Line Monitoring during Yeast Fermentations Based on Raman Spectroscopy. *Vib. Spectrosc.* **2024**, *132*, 103672. [CrossRef]
22. Feng, Y.; Tian, X.; Chen, Y.; Wang, Z.; Xia, J.; Qian, J.; Zhuang, Y.; Chu, J. Real-Time and on-Line Monitoring of Ethanol Fermentation Process by Viable Cell Sensor and Electronic Nose. *Bioresour. Bioprocess.* **2021**, *8*, 1–10. [CrossRef] [PubMed]

23. Kwon, H.; Shiu, J.; Rivera, E.C.; Yamakaya, C. Soft Sensor for Ethanol Fermentation Monitoring through Data-Driven Modeling and Synthetic Data Generation. In Proceedings of the 4th International Electronic Conference on Biosensors, Online, 20–22 May 2024; Volume 104, p. 30. [CrossRef]
24. Zhang, T.; Tian, X.; Liu, X.C.; Ye, J.A.; Fu, F.; Shi, X.T.; Liu, R.G.; Xu, C.H. Advances of Deep Learning in Electrical Impedance Tomography Image Reconstruction. *Front. Bioeng. Biotechnol.* **2022**, *10*, 1019531. [CrossRef]
25. Tinôco, D.; Batista Da Silveira, W. Kinetic Model of Ethanol Inhibition for *Kluyveromyces marxianus* CCT 7735 (UFV-3) Based on the Modified Monod Model by Ghose & Tyagi. *Biologia* **2021**, *76*, 3511–3519. [CrossRef]
26. Rzasa, M.; Łukasiewicz, E.; Wójtowicz, D. Test of a New Low-Speed Compressed Air Engine for Energy Recovery. *Energies* **2021**, *14*, 1179. [CrossRef]
27. Urry, L.A.; Cain, M.L.; Wasserman, S.A.; Minorsky, P.V.; Orr, R. *Biologia Campbella*, 12th ed.; REBIS: Poznań, Poland, 2025.
28. Yadav, M.; Yadav, H.S. *Biochemistry: Fundamentals and Bioenergetics*; Bentham Science Publisher: Sharjah, United Arab Emirates, 2021; Volume 418.
29. Ji, X.X.; Zhang, Q.; Yang, B.X.; Song, Q.R.; Sun, Z.Y.; Xie, C.Y.; Tang, Y.Q. Response Mechanism of Ethanol-Tolerant *Saccharomyces cerevisiae* Strain ES-42 to Increased Ethanol during Continuous Ethanol Fermentation. *Microb. Cell Fact.* **2025**, *24*, 1–15. [CrossRef]
30. Wolf, I.R.; Marques, L.F.; de Almeida, L.F.; Lázari, L.C.; de Moraes, L.N.; Cardoso, L.H.; Alves, C.C.d.O.; Nakajima, R.T.; Schnepper, A.P.; Golim, M.d.A.; et al. Integrative Analysis of the Ethanol Tolerance of *Saccharomyces cerevisiae*. *Int. J. Mol. Sci.* **2023**, *24*, 5646. [CrossRef]
31. Christofi, S.; Papanikolaou, S.; Dimopoulou, M.; Terpou, A.; Cioroiu, I.B.; Cotea, V.; Kallithraka, S. Effect of Yeast Assimilable Nitrogen Content on Fermentation Kinetics, Wine Chemical Composition and Sensory Character in the Production of Assyrtiko Wines. *Appl. Sci.* **2022**, *12*, 1405. [CrossRef]
32. Tse, T.J.; Wiens, D.J.; Reaney, M.J.T. Production of Bioethanol—A Review of Factors Affecting Ethanol Yield. *Fermentation* **2021**, *7*, 268. [CrossRef]
33. *IEEE Std 802.3-2002 (Revision of IEEE Std 802.3, 2000 edn)*; 802.3-2022—IEEE Standard for Information technology-Telecommunications and Information Exchange Between Systems-Local and Metropolitan Area Networks-Specific Requirements Part 3: Carrier Sense Multiple Access with Collision Detection (CSMA/CD) Access Method and Physical Layer Specifications. IEEE: New York, NY, USA, 2002; pp. 1–1550. [CrossRef]
34. Konopacka, A.; Konopacki, M.; Kordas, M.; Rakoczy, R. Mathematical Modeling of Ethanol Production by *Saccharomyces cerevisiae* in Batch Culture with Non-Structured Model. *Chem. Process Eng.* **2023**, *40*, 281–291. [CrossRef]
35. Kulisz, M.; Kłosowski, G.; Rymarczyk, T.; Słoniec, J.; Gauda, K.; Cwynar, W. Optimizing the Neural Network Loss Function in Electrical Tomography to Increase Energy Efficiency in Industrial Reactors. *Energies* **2024**, *17*, 681. [CrossRef]
36. Wang, J.; Deng, J.; Liu, D. Deep Prior Embedding Method for Electrical Impedance Tomography. *Neural Netw.* **2025**, *188*, 107419. [CrossRef]
37. Slouka, C.; Wurm, D.J.; Brunauer, G.; Welzl-Wachter, A.; Spadiut, O.; Fleig, J.; Herwig, C. A Novel Application for Low Frequency Electrochemical Impedance Spectroscopy as an Online Process Monitoring Tool for Viable Cell Concentrations. *Sensors* **2016**, *16*, 1900. [CrossRef] [PubMed]
38. Díaz Pacheco, A.; Delgado-Macuil, R.J.; Larralde-Corona, C.P.; Dinorín-Téllez-Girón, J.; Martínez Montes, F.; Martinez Tolibia, S.E.; López y López, V.E. Two-Methods Approach to Follow up Biomass by Impedance Spectroscopy: Bacillus Thuringiensis Fermentations as a Study Model. *Appl. Microbiol. Biotechnol.* **2022**, *106*, 1097–1112. [CrossRef]
39. Díaz Pacheco, A.; Dinorín-Téllez-Girón, J.; Martínez Montes, F.J.; Martínez Tolibia, S.E.; López y López, V.E. On-Line Monitoring of Industrial Interest Bacillus Fermentations, Using Impedance Spectroscopy. *J. Biotechnol.* **2022**, *343*, 52–61. [CrossRef]
40. Salakkam, A.; Phukoetphim, N.; Laopaiboon, P.; Laopaiboon, L. Mathematical Modeling of Bioethanol Production from Sweet Sorghum Juice under High Gravity Fermentation: Applicability of Monod-Based, Logistic, Modified Gompertz and Weibull Models. *Electron. J. Biotechnol.* **2023**, *64*, 18–26. [CrossRef]
41. Veloso, I.I.K.; Rodrigues, K.C.S.; Batista, G.; Cruz, A.J.G.; Badino, A.C. Mathematical Modeling of Fed-Batch Ethanol Fermentation Under Very High Gravity and High Cell Density at Different Temperatures. *Appl. Biochem. Biotechnol.* **2022**, *194*, 2632–2649. [CrossRef] [PubMed]
42. Lairón-Peris, M.; Routledge, S.J.; Linney, J.A.; Alonso-del-Real, J.; Spickett, C.M.; Pitt, A.R.; Guillamón, J.M.; Barrio, E.; Goddard, A.D.; Querol, A. Lipid Composition Analysis Reveals Mechanisms of Ethanol Tolerance in the Model Yeast *Saccharomyces cerevisiae*. *Appl. Environ. Microbiol.* **2021**, *87*, 1–22. [CrossRef] [PubMed]
43. Sahana, G.R.; Balasubramanian, B.; Joseph, K.S.; Pappuswamy, M.; Liu, W.C.; Meyyazhagan, A.; Kamyab, H.; Chelliapan, S.; Joseph, B.V. A Review on Ethanol Tolerance Mechanisms in Yeast: Current Knowledge in Biotechnological Applications and Future Directions. *Process Biochem.* **2024**, *138*, 1–13. [CrossRef]

44. Cairns, P.; Hamilton, L.; Racine, K.; Phetxumphou, K.; Ma, S.; Lahne, J.; Gallagher, D.; Huang, H.; Moore, A.N.; Stewart, A.C. The Science of Beer Effects of Hydroxycinnamates and Exogenous Yeast Assimilable Nitrogen on Cider Aroma and Fermentation Performance Effects of Hydroxycinnamates and Exogenous Yeast Assimilable Nitrogen on Cider Aroma and Fermentation Performance. *J. Am. Soc. Brew. Chem.* **2021**, *2022*, 236–247. [CrossRef]
45. Godillot, J.; Sanchez, I.; Perez, M.; Picou, C.; Galeote, V.; Sablayrolles, J.M.; Farines, V.; Mouret, J.R. The Timing of Nitrogen Addition Impacts Yeast Genes Expression and the Production of Aroma Compounds During Wine Fermentation. *Front. Microbiol.* **2022**, *13*, 829786. [CrossRef]
46. Li, J.; Yuan, M.; Meng, N.; Li, H.; Sun, J.; Sun, B. Influence of Nitrogen Status on Fermentation Performances of Non-Saccharomyces Yeasts: A Review. *Food Sci. Hum. Wellness* **2024**, *13*, 556–567. [CrossRef]
47. Golmohamadi, H. Demand-Side Management in Industrial Sector: A Review of Heavy Industries. *Renew. Sustain. Energy Rev.* **2022**, *156*, 111963. [CrossRef]
48. Sharma, D.; Singh, K. AI-Enhanced Bioprocess Technologies: Machine Learning Implementations from Upstream to Downstream Operations. *World J. Microbiol. Biotechnol.* **2025**, *41*, 1–29. [CrossRef] [PubMed]
49. Dotzauer, M. Determining Optimal Component Configurations for Flexible Biogas Plants Based on Power Prices of 2020–2022 and the Legislation Framework in Germany. *Renew Energy* **2024**, *236*, 121252. [CrossRef]
50. Stouffer, K.; Pease, M.; Tang, C.; Zimmerman, T.; Pillitteri, V.; Lightman, S.; Hahn, A.; Saravia, S.; Sherule, A.; Thompson, M. Guide to Operational Technology (OT) Security. 2023. Available online: https://nvlpubs.nist.gov/nistpubs/SpecialPublications/NIST.SP.800-82r3.pdf (accessed on 6 October 2025). [CrossRef]

energies

Article

Data-Driven Based Dynamic State Estimation Method for Regional Integrated Energy Systems Incorporating Multi-Dimensional Generation-Grid-Load Characteristics

Shengwen Li [1], Xiao Chang [1], Liang Ji [2,*] and Junchen Mao [2]

[1] State Grid Shanxi Electric Power Company Electric Power Research Institute, Taiyuan 030001, China; m7876277@163.com (S.L.); x565652025@163.com (X.C.)
[2] Shanghai University of Electric Power, Shanghai 201399, China; jls8899564@163.com
* Correspondence: jiliang@shiep.edu.cn

Abstract

The regional integrated energy system (RIES) has emerged as a critical focus in energy systems research. The comprehensive incorporation of renewable energy and inherent multi-energy flow interconnection within RIES markedly elevates the complexity of "generation-load" balance regulation. Traditional model-driven dynamic state estimation methods, however, are constrained by fundamental limitations—complex modeling, inadequate representation of multi-energy flow interdependencies, and poor computational efficiency. This study proposes a data-driven dynamic state estimation method for RIES, utilizing multi-dimensional "generation-grid-load" characteristic information as its primary input and employing a synergistic framework of Empirical Mode Decomposition-Singular Value Decomposition (EMD-SVD) alongside an enhanced Bidirectional Long Short-Term Memory (BiLSTM) network. EMD-SVD preprocesses raw data to remove noise and extract essential features, while the enhanced BiLSTM serves a dual purpose: it first attains high-precision photovoltaic output prediction and multi-energy load forecasting and subsequently evaluates the node states of the multi-energy flow coupling system. A case study on a practical coupled RIES, comprising a 33-node power system, 7-node gas system, and 6-node thermal system, demonstrates that the proposed method achieves high estimation accuracy and remarkable computational efficiency while effectively addressing the inherent limitations of conventional model-driven approaches.

Keywords: regional integrated energy system; multi-energy flow coupling; data-driven; multi-energy load forecasting; dynamic state estimation approach

1. Introduction

In light of the rapid global energy transition and increasing environmental concerns, it is essential to build renewable energy systems that ensure energy security, economic feasibility, and sustainable development. The RIES serves as a crucial physical conduit for attaining carbon peaking and carbon neutrality objectives by dismantling obstacles between conventional single-energy systems. It facilitates multi-energy complementarity, cascaded energy usage, and integrated management of "generation-grid-load-storage," thereby greatly enhancing energy utilization efficiency, promoting the absorption of clean energy and efficiently reducing carbon emissions. Consequently, RIES has emerged as a central emphasis of energy policy and technical research on both national and global scales [1–5]. To attain precise perception, real-time monitoring, and effective energy management of RIES, while facilitating its coordinated dispatching and informed decision-making, it is essential to derive optimal estimated values of the system's real-time operational status via dynamic state estimation technology [6,7].

State estimation for integrated energy systems is categorized into static and dynamic state estimation. Static state estimation methods provide a reliable basis for evaluating the operational status of integrated energy systems in steady-state conditions. Researchers in [8] delineated the integrated electricity-heating network by partitioning the heating network into a hydraulic model and a thermal model. Utilizing a semi-joint static model, they employed the Weighted Least Squares (WLS) approach to perform a joint state estimate for the electricity-heating network. In [9], the WLS approach was utilized for state estimation in the electricity-gas integrated energy system under incomplete measurement settings. In [10], the Weighted Least Absolute Value (WLAV) method was employed for state assessment of electricity-gas networks; experiments with various forms of erroneous data demonstrated that this method exhibits considerable robustness. Researchers in [11] suggested a Bilinear Weighted Least Absolute Value (BWLAV) method derived from WLAV, which effectively addresses the issue of initial value configuration for gas networks and enhances computing efficiency. A three-stage state estimate technique for integrated energy systems comprising distributed electricity, heating, and gas, utilizing the Alternating Direction technique of Multipliers (ADMM), was proposed in [12]. This approach circumvents the non-convex optimization dilemma arising from nonlinear measurements and enhances the model's convergence. In [13], state estimation methods for combined gas-electricity networks were explored; a coupling mechanism characterization framework and practical analysis tools were established to lay a theoretical foundation for state estimation optimization in coordinated operation. In [14], gas transients were incorporated into the state estimation of integrated energy systems; gas flow partial differential equations were simplified via discretization criteria, dynamic modeling was optimized, and its accuracy in real-time state assessment outperforms traditional methods. All the aforementioned studies concentrate on the static state estimation of integrated energy systems. Nonetheless, the extensive incorporation of new energy into power grids, along with the pronounced disparities in dynamic operational characteristics among subsystems within RIES, has markedly amplified its volatility and randomness. In a complex and dynamic context, a singular static state assessment is unlikely to provide correct situational knowledge of RIES. Consequently, it is essential to implement dynamic state estimation as an adjunct.

At present, research on dynamic state estimation for RIES is in the developmental phase. Researchers in [15] utilized the node method to simulate the dynamic processes of heating systems and subsequently devised a dynamic state estimation technique appropriate for integrated electricity-heating networks. In [16], the dynamic properties of gas transmission were elucidated via the discretization of natural gas pipeline equations, and a multi-time-scale dynamic estimation framework for integrated electricity-gas energy systems was established. Despite the consideration of dynamic properties in gas and heating networks by these two dynamic state estimation methods, their fundamental algorithms remain dependent on the conventional WLS technique, resulting in inadequate accuracy in monitoring and forecasting dynamic processes. In [17], researchers adeptly employed the unified multi-energy flow theory by transforming time-domain partial differential equations into linear algebraic equations via frequency-domain transformation, thereby constructing a dynamic state estimation model for natural gas systems. In [18], to tackle the multi-time-scale challenge of electricity-gas-heating systems, the nonlinear dynamic model was converted into an algebraic format utilizing the finite element approach, and Kalman Filtering was employed for dynamic state estimation. Nonetheless, the problem of elevated computational complexity persists and requires resolution. The aforementioned works have significantly advanced model-driven state estimates for integrated energy systems. Nevertheless, the intricate computations of dynamic processes in gas and heating networks compromise the accuracy and real-time efficacy of model-driven methodologies. Moreover, gas and heating networks typically lack measurement instruments, and the terminal systems are intricate. These characteristics complicate the acquisition of real-time measurement data for thermal and gas loads, posing significant obstacles to the state estimation of RIES.

This study focuses on "generation-grid-load", extracts multi-dimensional features, uses data-driven technology to tackle the identified challenges, and performs dynamic state estimation for RIES. EMD-SVD functions as a filtering module for processing raw meteorological data, while the BiLSTM network serves as the framework for forecasting "generation-load" status. During the offline training phase, it utilizes highly correlated feature data to predict photovoltaic power output and historical data to predict node loads, ultimately relying on BiLSTM to execute the dynamic state estimation of the multi-energy flow coupling system in RIES.

2. Algorithm Architecture of Dynamic State Estimation Integrating Multi-Dimensional "Generation-Grid-Load" Features

Figure 1 illustrates the conceptual diagram of the algorithm presented in this research, with the blue section denoting the natural gas system, the red section indicating the heating system, and the black section representing the electricity system. This conceptual diagram encompasses the data input and state estimation model of the electricity-heat-gas network: On the "generation" side, the system achieves high-precision photovoltaic power forecasting by integrating historical photovoltaic power generation data with critical time-series meteorological feature data; on the "load" side, it amalgamates time-series meteorological feature data and historical node load data for load forecasting; and based on the established "generation-load" coordinated state benchmark, it facilitates dynamic state estimation of the electricity-heat-gas coupled RIES on the "grid" side.

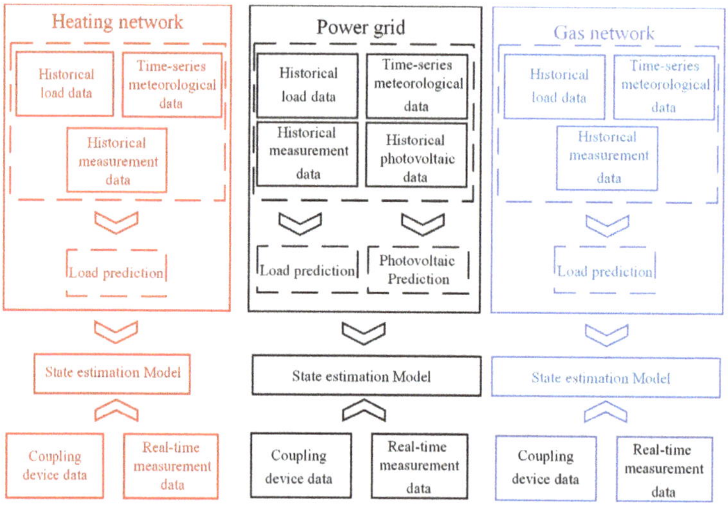

Figure 1. Algorithm concept diagram.

3. Generation-Load Prediction Based on EMD-SVD-BiLSTM

3.1. EMD Decomposition

EMD is a technique that disaggregates data according to its distinctive time scales to derive various intrinsic mode functions (IMFs). Various IMF components signify unique characteristic fluctuation sequences, and the decomposed IMFs can elucidate the fluctuation properties of the original data across multiple time scales. The precise decomposition procedure is outlined as follows:

(1) Take an original data sequence $x_0(t)$. Find all maximum value points in the sequence. Make an upper envelope with cubic spline interpolation. Find all minimum value points. Make a lower envelope with cubic spline interpolation. Let $m_0(t)$ be the mean of the upper and lower envelopes. Then, the first decomposed component $h_0(t)$ is $h_1(t) = x_0(t) - m_0(t)$. The remaining component is $r_1(t) = x_0(t) - h_1(t)$.

(2) When decomposing the n-th$(n \geq 2)$ component, take $h_{n-1}(t)$ as the original data. Make upper and lower envelopes with cubic spline interpolation between the maximum and minimum value points. Calculate the mean $m_{n-1}(t)$. Then, the decomposed vector is $h_n(t) = x_{n-1}(t) - m_{n-1}(t)$. The remaining component $r_n(t) = x_0(t) - \sum_{k=1}^{n} h_k(t)$.

(3) Continue Step 2 until the maximum number of decomposition iterations is attained, or the resultant component from decomposition is a monotonic function; subsequently, conclude the iteration process.

(4) The EMD decomposition presented in this paper is utilized to decompose original data for photovoltaic power and load prediction. The upper and lower envelopes are generated using data from the time period corresponding to the input of the BiLSTM neural network during the decomposition process, followed by the actual decomposition.

3.2. SVD Data Dimensionality Reduction

SVD is employed to decrease data dimensionality. It can eliminate noise in data and retrieve the majority of the feature information. The preliminary data decomposed by EMD, characterized by numerous sequence dimensions and inconsistent data scales, when directly utilized as input for BiLSTM, may result in certain information being unresponsive to neuronal adjustments during error backpropagation and weight updates, thereby diminishing the predictive efficacy of the neural network.

Let SVD decompose the input matrix A into singular values and corresponding subspaces. It is expressed as follows:

$$\boldsymbol{A}=\boldsymbol{U}\boldsymbol{S}\boldsymbol{V}^T \tag{1}$$

Here, $\boldsymbol{U}$ is the left singular matrix. It is an orthogonal basis for the column vector space of input matrix A; $\boldsymbol{S}$ is the singular value matrix. It has singular values sorted from largest to smallest. Matrix $\boldsymbol{V}$ is the right singular matrix. It is an orthogonal basis for the row vector space of input matrix A.

To decompose A into $\boldsymbol{USV}^T$ and use SVD to filter original data, follow the following steps:

(1) Calculate the eigenvalues and eigenvectors of $\boldsymbol{AA}^T$ and $\boldsymbol{A}^T\boldsymbol{A}$, Then sort these eigenvalues and eigenvectors from largest to smallest by eigenvalues.

(2) Form left singular matrix $\boldsymbol{U}$. It is made of the eigenvector matrix of $\boldsymbol{AA}^T$ after sorting from largest to smallest. Its singular value matrix is diagonal. The diagonal elements are square roots of the sorted eigenvalues of $\boldsymbol{AA}^T$

(3) Form right singular matrix $\boldsymbol{V}$. It is made of the eigenvector matrix of $\boldsymbol{A}^T\boldsymbol{A}$ after sorting from largest to smallest.

(4) When using SVD to reduce the dimension of multi-dimensional data, if the required dimension is $\boldsymbol{k}$:

$$A_k = \boldsymbol{U}_k S_k \tag{2}$$

A_k is the data after dimensionality reduction. $\boldsymbol{U}_k$ is the first $\boldsymbol{k}$ column of the left singular matrix $\boldsymbol{U}$. S_k is the first $\boldsymbol{k}$ dimensional data of the singular value matrix $\boldsymbol{S}$.

The application of the EMD-SVD method to decompose environmental elements in photovoltaic power and load prediction mitigates the issue of data singularity in prediction outcomes resulting from too intricate input vector information. This procedure aligns more closely with the methodology of this study, which employs projected "generation-load" states for the dynamic state estimate of RIES.

Figure 2 illustrates the sequence diagram of solar irradiance data for photovoltaic power forecasting subsequent to EMD-SVD processing and normalization. "OI" denotes the sequence of original data post-normalization, while "PI1–PI6" signify sequences obtained following EMD-SVD decomposition and filtering, utilized as inputs for the BiLSTM network.

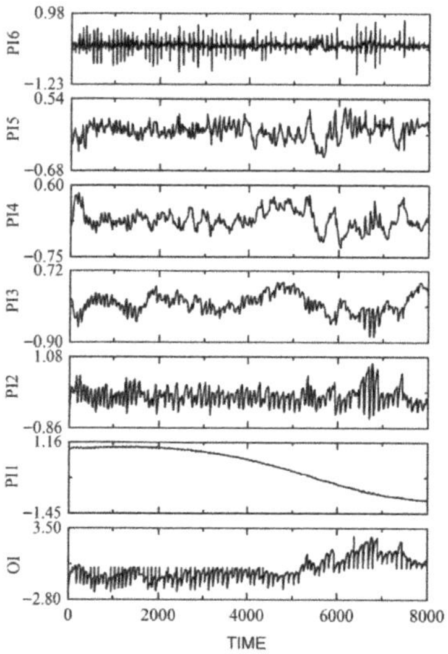

Figure 2. EMD-SVD decomposition sequence.

3.3. *EMD-SVD-BiLSTM Model*

Following the execution of EMD-SVD decomposition on the original dataset, the resultant data have extracted essential information and eliminated certain singular data points. This research proposes a system that uses BiLSTM as the primary framework for solar power and load prediction, utilizing input data processed by EMD-SVD.

BiLSTM is derived from RNN [19] and LSTM [20], which are engineered to manage data exhibiting temporal correlations in sequences. Recurrent Neural Networks (RNNs) encounter the issue of gradient vanishing when processing long-term sequences, but Long Short-Term Memory (LSTM) networks mitigate this problem by incorporating a state mechanism that retains long-term dependencies. BiLSTM represents an enhanced version of unidirectional LSTM. BiLSTM integrates a forward LSTM layer with a backward LSTM layer, generating output by utilizing both layers concurrently in opposite directions [21]. BiLSTM can optimize the utilization of RIES historical data and comprehensively account for both forward and backward information, hence enhancing predictive performance and accuracy.

Figure 3 shows a typical structure diagram of BiLSTM. $\boldsymbol{X_{t-1}}, \boldsymbol{X_t}, \boldsymbol{X_{t+1}}$ stand for the input data at each moment. $\boldsymbol{LSTM_F}$ and $\boldsymbol{LSTM_B}$ stand for the hidden layer states of the forward LSTM layer and backward LSTM network layer, respectively. $\boldsymbol{O_{t-1}}, \boldsymbol{O_t}, \boldsymbol{O_{t+1}}$ stand for the corresponding output data. $\omega_1, \omega_2, \ldots, \omega_6$ stand for the weight transmission between different layers. The update states of the forward LSTM, backward hidden layer, and the output state of BiLSTM are shown in Formulas (3)–(5):

$$\boldsymbol{LSTM_{F,i}} = f_1(\omega_1 \boldsymbol{X_i} + \omega_4 \boldsymbol{LSTM_{F,i-1}}) \tag{3}$$

$$\boldsymbol{LSTM_{B,i}} = f_1(\omega_2 \boldsymbol{X_i} + \omega_3 \boldsymbol{LSTM_{B,i-1}}) \tag{4}$$

$$\boldsymbol{O_i} = f_1(\omega_5 \boldsymbol{LSTM_{F,i}} + \omega_6 \boldsymbol{LSTM_{B,i}}) \tag{5}$$

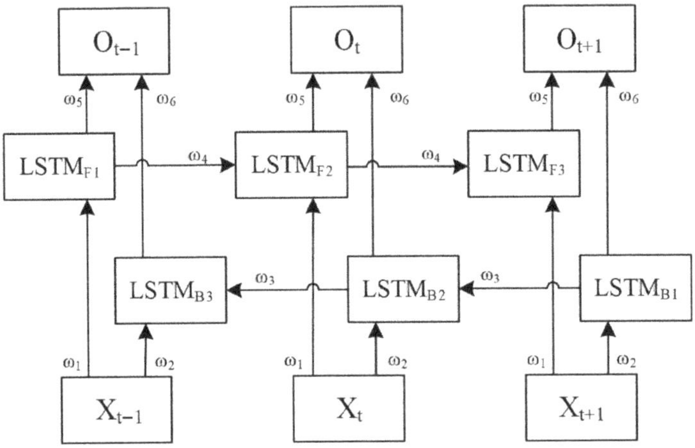

Figure 3. Framework diagram of BiLSTM.

In the formulas, f_1, f_2, f_3 are activation functions between different layers, and i is the corresponding moment when the output responds.

The method presented in this research utilizes a temporal window format for the input during the training and application of BiLSTM, as illustrated in Figure 4. The normalized data from the time window preceding and succeeding the prediction instant is utilized as the input vector, with the time window advancing incrementally as time elapses, so using the attributes of BiLSTM.

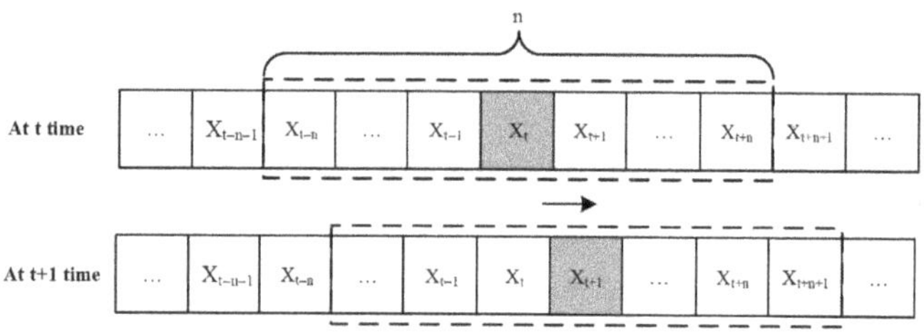

Figure 4. Schematic diagram of the BiLSTM time window.

3.4. Photovoltaic Prediction Based on EMD-SVD-BiLSTM

Deep learning is extensively employed for modeling photovoltaic power prediction because of its capability to effectively represent the relationship between meteorological variables and solar power output, as well as its adaptability to fluctuations. The literature [22] suggests that incorporating more pertinent climatic variables will significantly enhance the accuracy of photovoltaic power projection. In research [23], the Pearson correlation analysis is typically employed to identify the input variables for the solar power prediction model and to extract highly correlated feature data. It uses the symbol ρ to reflect the degree of linear correlation between two variables, where ρ ranges from -1 to 1. If $\rho > 0$, it indicates that the two variables are positively correlated; if $\rho < 0$, they are negatively correlated. The closer the absolute value of ρ is to 1, the stronger the correlation between the two variables. Its calculation formula is as follows:

$$\rho = \frac{\sum_{i=1}^{n} (x_i - \overline{X})(Y_i - \overline{Y})}{\sqrt{\sum_{i=1}^{n} (X_i - \overline{X})^2}\sqrt{\sum_{i=1}^{n} (Y_i - \overline{Y})^2}} \tag{6}$$

where X and Y represent the values of the two variables, respectively; $\overline{X}$ and $\overline{Y}$ represent the sample means of the two variables, respectively; and n denotes the number of samples.

Table 1 shows the Pearson correlation analysis of the input vector for photovoltaic power output prediction. For the photovoltaic power output prediction in this paper, the selected input variables are: local solar irradiance G, air temperature T_a, and humidity RH; the output is the real-time power output of the photovoltaic power station, i.e.,

$$\begin{aligned} Input_{PV} &= \{G, T_a, RH\} \\ Output_{PV,i} &= \{P_{PV,i}\} \end{aligned} \tag{7}$$

Table 1. Pearson coefficient analysis table of photovoltaic forecast.

Feature	Correlation Coefficient
Temperature	0.763
Solar irradiance	0.979
Humidity	0.338
Wind speed	0.187

3.5. Load Forecasting Model Based on EMD-SVD-BiLSTM

Load forecast resembles photovoltaic power prediction due to its significant link with meteorological parameters. The multi-component loads of the RIES are intricately linked to human lifestyle and production activities, hence exhibiting significant temporal association.

Load prediction, while dependent on meteorological data, demonstrates distinct periodicity and characteristics, influenced by seasonal fluctuations, temperature variations, and the evolving patterns of industrial operations and daily activities of people. Utilizing the BILSTM mechanism, which links data in a temporal sequence, enhances the effects of specific dates and seasonal variations in periodic load prediction, thereby increasing the accuracy of load forecasting. Table 2 displays the Pearson correlation analysis of the input vector for multi-component load forecasting.

Table 2. Pearson coefficient analysis table for load forecasting.

Feature	Electrical Load	Thermal Load	Gas Load
Hour	0.7243	0.6852	0.5937
Date	0.0263	0.0315	0.0289
Month	−0.0053	−0.0121	−0.0086
Temperature	0.4402	0.7965	0.6538
Humidity	−0.2874	−0.4328	0.4876
Wind speed	0.1674	0.0247	−0.0289
Air pressure	−0.5387	−0.5846	0.0658

Table 2 illustrates large variations in the response intensity of multi-component loads to identical climatic influencing factors.

For electrical load prediction, by combining feature correlation and demand logic, select air temperature T_a, humidity RH, air pressure P_C, date, and specific hour of the current moment as inputs.

Thermal load is affected by meteorological factors like temperature and air pressure, and time-dimension features. So, select temperature T_a, humidity RH, air pressure P_C, date, and hour. Temperature leads thermal demand change; air pressure relates to heat transfer and building energy consumption environment; and time dimension anchors the thermal load's periodic fluctuation. They form a multi-dimensional feature space together.

Gas load is affected by temperature, humidity, wind speed, and time features affect gas usage patterns. With the Pearson coefficient, select temperature T_a, humidity RH, wind speed V_w, date, and hour. Temperature relates directly to gas-fired heating demand; humidity and wind speed affect equipment efficiency and user habits; and time dimension ensures accurate depiction of gas usage cycles.

In summary, the input–output relationships for electrical, thermal, and gas load predictions are as follows:

$$\begin{cases} Input_{load,elec} = \{T_a, RH, P_C, Date, Hour\} \\ Output_{load,elec} = \{Load_{elec}\} \end{cases} \tag{8}$$

$$\begin{cases} Input_{load,heat} = \{T_a, RH, P_C, Date, Hour\} \\ Output_{load,heat} = \{Load_{heat}\} \end{cases} \tag{9}$$

$$\begin{cases} Input_{load,gas} = \{T_a, RH, V_w, Date, Hour\} \\ Output_{load,gas} = \{Load_{gas}\} \end{cases} \tag{10}$$

4. Dynamic State Estimation Model of RIES Based on "Generation-Load" Status

This article initially employs the EMD-SVD-BiLSTM methodology to forecast the solar power output on the power grid side within RIES to ascertain the "generation" state. Subsequently, multi-component load forecasting is conducted on the pertinent nodes of the electricity grid, heating network, and gas network to ascertain the "load" state. Dynamic state estimation is thereafter conducted on the internal nodes of RIES, based on the assessment of the "generation-load" status.

4.1. Algorithm Flow Chart

Figure 5 illustrates the flowchart of the dynamic state estimation model of RIES predicated on the "generation-load" condition. The "generation-load" balance condition of RIES at future intervals is ascertained by predictive approaches, followed by the execution of a dynamic state estimate of RIES.

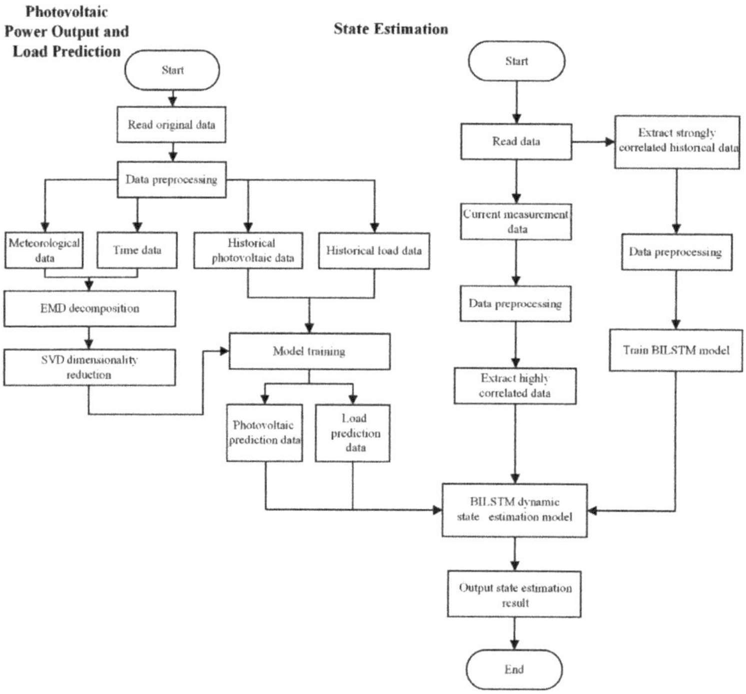

Figure 5. Algorithm flow chart.

4.2. Dynamic State Estimation Model of Coupled Nodes

This study utilizes data from RIES coupling devices to train the model for coupled nodes in RIES, with the objective of enhancing state estimation performance for these nodes. The coupling devices utilized in this context are gas turbines (GT) and Combined Heat and Power (CHP) units.

A gas turbine is a standard component for electricity-gas connection. It utilizes gas generated from natural gas combustion to propel itself and perform labor, hence producing electricity. The energy conversion relationship of a gas turbine is articulated as follows:

$$P_{\mathrm{GT}} = \frac{1}{3600}\eta_{\mathrm{GT}} G_{\mathrm{GT}} q_{\mathrm{ng}} \tag{11}$$

In the formula, P_{GT} represents the power output of the gas turbine, measured in MW. η_{GT} is the power generation efficiency of the gas turbine (a dimensionless quantity). G_{GT}

denotes the gas consumption of the gas turbine, with the unit of Nm3/h. q_{ng} stands for the calorific value of natural gas, and its unit is MJ/Nm3.

CHP is among the most prevalent coupling units in electricity, heat, and gas networks. In CHP systems utilizing gas turbines or reciprocating internal combustion engines as prime movers, the heat production power ratio is regarded as constant. The electro-thermal ratio is articulated as:

$$\gamma_{CHP} = Q_{CHP}/P_{CHP} \tag{12}$$

In the formula: γ_{chp} represents the electro-thermal ratio of CHP; Q_{CHP} is the heat production power of CHP; and P_{CHP} denotes the electricity generation power of the CHP unit.

This paper employs a BiLSTM neural network to develop distinct state estimation models for CHP units and GT units individually. The model for CHP units utilizes the unit's operational parameters and temperature data from heat supply ports as input, aiming to predict heat power as the output target. For GT units, the input consists of the unit's parameters and system pressure measurement data, with the output aimed at electricity generation power.

The intentional choice of these input and output characteristics is based on a study of how energy changes from one form to another. The selected feature parameters exhibit significant physical correlations with the target variables, aligning with the principle of energy conservation and accurately reflecting the unit's operational characteristics, thus enhancing the model's generalization capability and state estimate precision.

4.3. Model Robustness Optimization and Data Preprocessing

This study employs a noise-added data augmentation technique during the training phase to enhance the model's robustness, acknowledging the unavoidable existence of measured erroneous data in real-world RIES. It randomly selects samples from the original dataset, introduces Gaussian white noise with a standard deviation of 10%, and replicates measurement abnormalities in actual systems. Training the model on noise-contaminated datasets markedly improves its fault tolerance to erroneous data during actual operation.

Due to the randomness and minimal prevalence of erroneous data in integrated energy systems, merely using flawed data for training complicates the acquisition of a model with substantial robustness. Initially, it conducts outlier detection via the quartile approach, then employs a modified Akima cubic interpolation algorithm for the intelligent substitution of erroneous data, and reintegrates the rectified results into the anomaly identification process for following measurements. To resolve the dimensional disparity in electricity-heat-gas multi-generation heterogeneous data, it standardizes the input characteristics by max–min normalization to achieve uniform dimensions.

5. Case Study

This study presents an integrated system comprising a 33-node electricity network, a 7-node natural gas network, and a 6-node thermal network, collectively referred to as the electricity-gas-heat RIES network. The topology is illustrated in Figure 6, and the system parameters are cited in Reference [24].

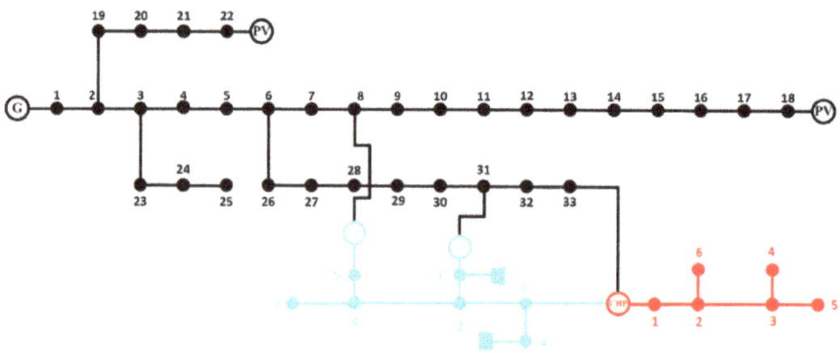

Figure 6. The electric-gas-heat coupling network.

Within the power network, nodes 18 and 22 are linked to photovoltaic devices; nodes 8 and 31 are associated with gas network nodes 5 and 1 through gas turbines; and node 33 is connected to gas network node 3 and thermal network node 1 via a CHP unit. Furthermore, gas network nodes 1 and 4 are linked to gas sources, while the thermal load demand of the thermal network is entirely met by the CHP unit.

The solar and load statistics were obtained from an integrated energy system in a Belgian park [25] and subsequently adjusted proportionally based on the RIES parameters utilized in this investigation. Both of these datasets have a sampling interval of 30 min; therefore, the time interval selected for the simulation process in this paper is also set to 30 min.

For the data analysis indicators of the simulation test system, this paper uses the Root Mean Square Error (RMSE) and the coefficient of determination R^2 to evaluate the accuracy of prediction and estimation.

$$RMSE(k) = \sqrt{\frac{1}{N}\sum_{i=1}^{N}(\hat{y}_{k,i} - y_{k,i})^2} \tag{13}$$

$$R^2(k) = 1 - \frac{\sum(y_{k,i} - \hat{y}_{k,i})^2}{\sum(y_{k,i} - \overline{y}_k)^2} \tag{14}$$

In the formulas, k is the node number, N is the length of the entire prediction sequence, $\hat{y}_{k,i}$ and $y_{k,i}$ are the predicted value and actual value of node k at time i, respectively, and $\overline{y}_k$ is the mean of the actual values for node k. RMSE measures the absolute error between predicted values and true values—a smaller RMSE means a smaller absolute error. Generally, R^2 ranges from 0 to 1; the closer it is to 1, the better the prediction performance. When R^2 is less than 0, the fitting of predicted values is even worse than just using the mean as the prediction.

5.1. Prediction Performance Analysis Based on EMD-SVD-Bilstm for "Generation-Load"

Figures 7 and 8 illustrate the outcomes of solar output prediction and multi-energy load prediction, respectively. In these pictures, the solar power station that is connected to Grid Node 22 is the prediction target.

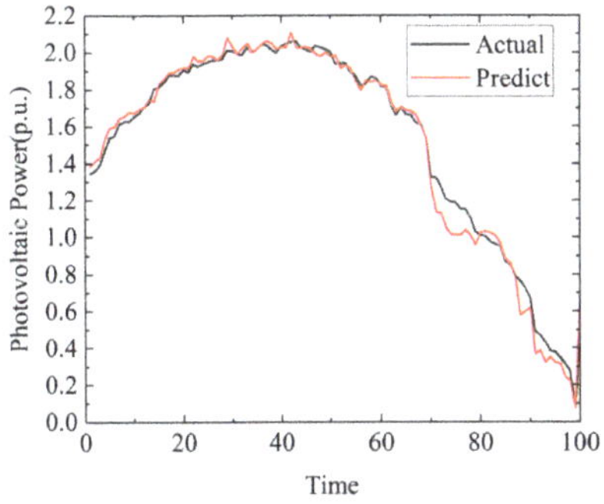

Figure 7. Photovoltaic power output prediction.

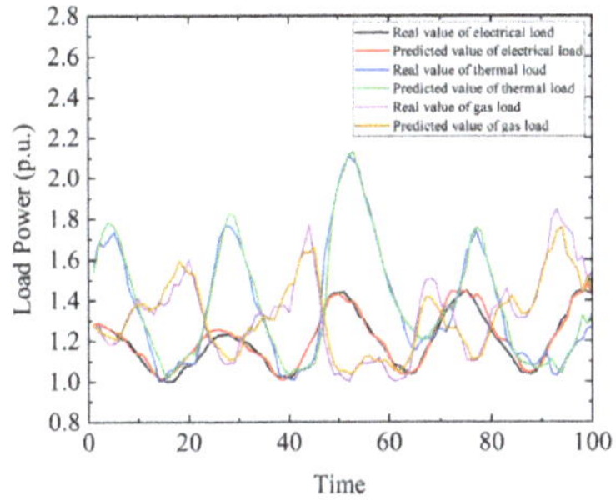

Figure 8. Multi-load forecasting.

For multi-energy load forecasting, Grid Node 16, Thermal Network Node 6, and Gas Network Node 3 in the RIES are selected as typical nodes to predict the loads of electricity, heat, and gas, respectively. All outcomes are presented in per unit (pu) value.

Table 3 displays the accuracy metrics for solar output prediction and multi-energy load forecasting.

Table 3. Prediction accuracy of PV output and multi-energy loads.

Prediction Type	RMSE	R^2	MAPE
Photovoltaic output	0.7709	0.9529	0.0425
Node electrical load	0.3642	0.9406	0.0538
Node thermal load	0.3825	0.9463	0.0754
Node gas load	0.5417	0.9587	0.0508

The data presented in the figures and Table 3 demonstrate that the "generation-load" prediction system employed in this study efficiently monitors fluctuations in photovoltaic output and load at RIES nodes. It is important to acknowledge that the predictions' accuracy diminishes when there are substantial temporal fluctuations in historical and meteorological data compared to a steady condition. Nonetheless, the overall predictive accuracy is commendable, and regarding load forecasting, this approach surpasses the method employing pseudo-measurements with significant noise to emulate multi-energy load measurements in the absence of real-time measurement setup.

5.2. Under "Generation-Load" Prediction: RIES Dynamic State Estimation

5.2.1. Result Analysis of Power Grid State Estimation

Figures 9 and 10 present comparative diagrams of the estimated voltage amplitude and phase angle of the power grid at Node 18, derived from calculations based on "generation-load" forecasting. Table 4 displays the RMSE and A values for the estimated power grid voltage vector.

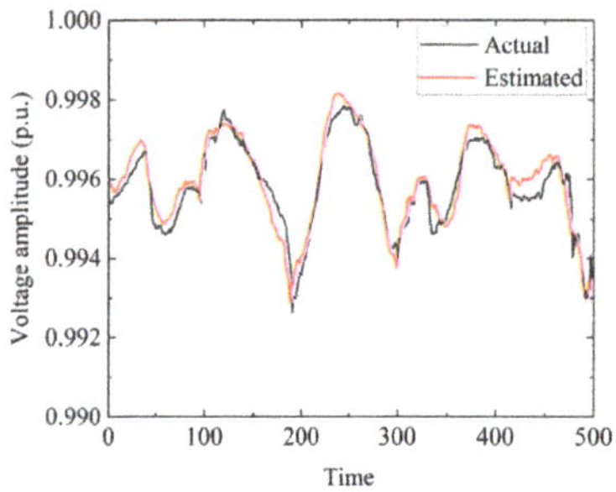

Figure 9. Estimated voltage amplitude of node 18.

Figure 11 shows the statistical R^2 diagram for voltage magnitude and voltage angle estimation of all nodes. The power grid is divided into four segments: Nodes 1–18, Nodes 19–22, Nodes 23–25, and Nodes 26–33. The R^2 estimation accuracy of the voltage vector for all nodes in the power grid is over 0.9. The segment of Nodes 19–22 is close to the main generator nodes and has photovoltaic devices. So, it is more stable than other nodes. This makes the estimated magnitude of the voltage vector closer to the actual value. In all, the method in this paper works well in the dynamic state estimation of the power grid.

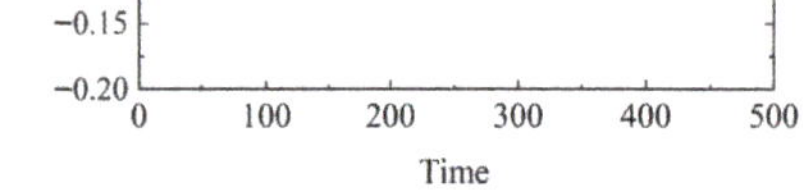

Figure 10. Estimated voltage angle of node 18.

Table 4. Table of RMSE and R^2 of voltage vector.

Node	RMSE		R^2		MAPE	
	Vm	Va	Vm	Va	Vm	Va
Node 10	0.0027	0.1062	0.9315	0.9313	0.0185	0.0191
Node 18	0.0004	0.0152	0.9461	0.9310	0.0126	0.0198
Average	0.0022	0.0735	0.9322	0.9324	0.0174	0.0174

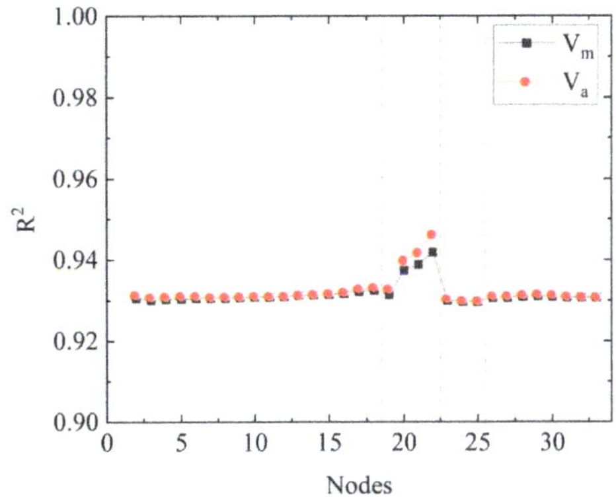

Figure 11. R^2 of estimated voltage vector.

5.2.2. Result Analysis of Thermal and Gas Network State Estimation

This paper aims to validate the efficacy of the proposed method for state estimation in thermal and gas networks by selecting Node 6 of the RIES thermal network and Node 3 of the gas network to conduct dynamic state estimation on the heating temperature of thermal network nodes and the pressure of gas network nodes, respectively, as illustrated in Figures 12 and 13. Tables 5 and 6 displays the RMSE and R^2 values for the predicted temperature of the thermal network nodes and the pressure of the gas network nodes, as well as the RMSE and R^2 values of the coupled nodes.

The dynamic state estimation results for the heating temperature at Thermal Network Node 6 and the pressure at Gas Network Node 3 indicate that the estimated curves closely align with the actual curves, albeit with slightly lower accuracy compared to the power grid state estimation. This is attributed to the influence of pipeline delays on data collecting in gas and heat networks. Nevertheless, in conjunction with the comparatively low RMSE values presented in the table and the coefficient of determination R^2 approaching 0.87, it can be inferred that the methodology proposed in this paper effectively characterizes the dynamic operating states of thermal and gas networks while demonstrating commendable state estimation performance.

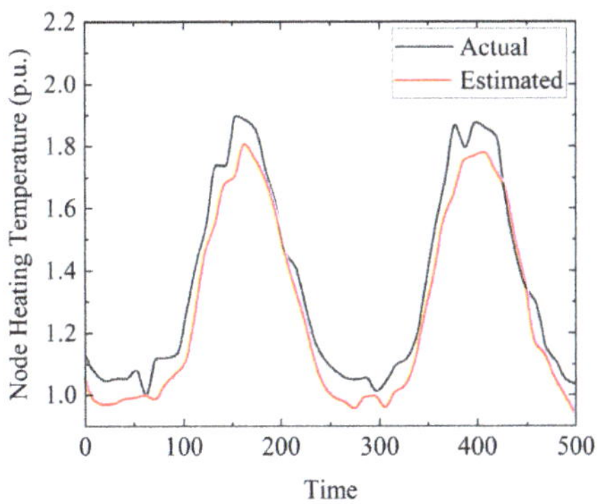

Figure 12. R^2 Heating temperature diagram of Node 6 in thermal network.

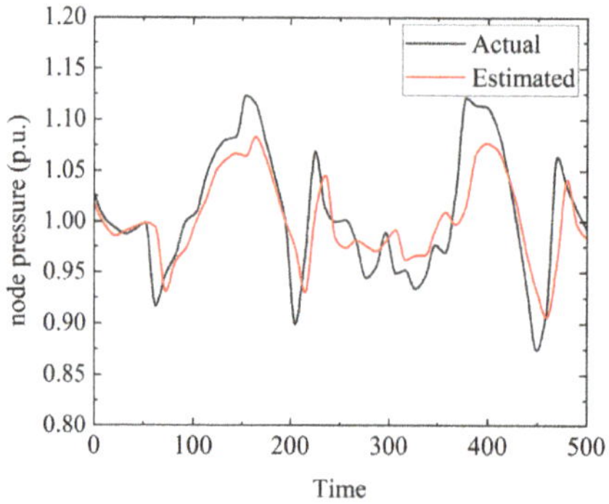

Figure 13. Pressure diagram of Node 3 in gas network.

Table 5. RMSE and R^2 of estimated thermal network node temperatures and gas network node pressures.

Nodes	RMSE	R^2	MAPE
Thermal Network Node 6	0.0624	0.8875	0.0624
Gas Network Node 3	0.0247	0.8763	0.0647
Average	0.0436	0.8864	0.0613

Table 6. RMSE and R^2 for coupled node estimation.

Node	RMSE	R^2	MAPE
Node 8	0.0137	0.0926	0.0461
Node 31	0.0176	0.8861	0.0611
Node 33	0.0253	0.8878	0.0607
Average	0.0185	0.8883	0.0594

6. Conclusions

This paper presents a data-driven dynamic state estimate approach for RIES that incorporates multi-dimensional attributes of "generation-grid-load." This method, addressing the practical operational conditions of RIES—marked by multi-energy coupling and extensive integration of renewable energy, resulting in a complex "generation-load" balance—successfully predicts photovoltaic output and multi-energy loads through EMD-SVD and a BiLSTM network, thereby establishing a dependable "generation-load" boundary for state estimation. Conversely, utilizing the anticipated "generation-load" states, the BILSTM network dynamically estimates the state variables of each node within the RIES multi-energy coupling system, circumventing the complexities of modeling and the challenges in accurately representing the multi-energy flow coupling characteristics inherent in traditional model-driven approaches. The case analysis results indicate that utilizing the RIES alongside a 33-node power system, a 7-node natural gas system, and a 6-node thermal system as the research subjects, the proposed method demonstrates high accuracy in estimating critical state variables, including thermal network node temperature and gas network node pressure, thereby validating the method's efficacy and computational efficiency advantages.

The efficacy of the suggested strategy depends on high-quality historical data and meteorological characteristic data. The experiment revealed that substantial temporal oscillations between these two data types result in a deterioration in the prediction accuracy of photovoltaic output and multi-energy loads. Consequently, due to pipeline transmission delays in gas and heat networks, the state estimate accuracy of their nodes is somewhat inferior to that of the power grid, unable to completely mitigate the measurement discrepancies induced by transmission delays. The experimental data are based on proportional changes related to a specific park-level RIES, and the model's responsiveness to RIES under varying climatic circumstances and energy architectures necessitates further validation. This paper fails to adequately address abrupt variations in renewable energy generation and multi-energy demand under harsh conditions, and the model's applicability to renewable energy integrated systems of different scales necessitates additional validation. In the future, two primary aspects of work will be prioritized: first, adopting an adaptive strategy to optimize model parameters online and improve resilience under extreme conditions; and second, expanding the technology's application in cross-regional interconnected RIES and validating its effectiveness in large-scale contexts.

Author Contributions: Conceptualization, S.L.; Methodology, X.C.; Investigation, L.J.; Writing—review and editing, J.M. All authors have read and agreed to the published version of the manuscript.

Funding: This work is supported by State Grid Shanxi Electric Power Company Science and Technology Project Research (52053024000V).

Data Availability Statement: The original contributions presented in this study are included in the article. Further inquiries can be directed to the corresponding author.

Conflicts of Interest: Author Shengwen Li, Xiao Chang were employed by the company State Grid Shanxi Electric Power Company Electric Power Research Institute. The remaining authors declare that the research was conducted in the absence of any commercial or financial relationships that could be construed as a potential conflict of interest.

Abbreviations

The following abbreviations are used in this manuscript:

RIES	Regional Integrated Energy System
EMD–SVD	Empirical Mode Decomposition-Singular Value Decomposition
BiLSTM	Bidirectional Long Short-Term Memory
RNN	Recurrent Neural Network
LSTM	Long Short-Term Memory
WLS	Weighted Least Square
WLAV	Weighted Least Absolute Value
BWLAV	Bilinear Weighted Least Absolute Value
ADMM	Alternating Direction Method of Multipliers
GT	Gas Turbines
CHP	Combined Heat and Power
RMSE	Root Mean Square Error
MAPE	Mean Absolute Percentage Error

References

1. Yuan, Y.; Miao, A.K.; Wu, H.; Zhu, J.P.; Wang, Z.M.; Qian, K. Review of the research framework and key issues for low-carbon integrated energy system. *High Volt. Eng.* **2024**, *50*, 4019–4036. [CrossRef]
2. Kang, C.Q.; Du, E.S.; Li, Y.W.; Zhang, N.; Chen, Q.X.; Guo, H.Y.; Wang, P. Key scientific problems and research framework for carbon perspective research of new power systems. *Power Syst. Technol.* **2022**, *46*, 821–833. [CrossRef]
3. Wang, Y.Z.; Kang, L.G.; Zhang, J.; Zhao, W.; Zhu, Y.L.; Zhao, J. Development history, typical form and future trend of integrated energy system. *Acta Energiae Solaris Sin.* **2021**, *42*, 84–95. [CrossRef]

4. Fan, J.B.; Meng, Z.J.; Shan, M.W.; Lü, L.J. A review of key technologies and international standardization of multi energy coupling systems. *Power Syst. Clean Energy* **2023**, *39*, 1–9.
5. Zhang, S.X.; Wang, D.Y.; Cheng, H.Z.; Song, Y.; Yuan, K.; Du, W. Key technologies and challenges of low-carbon integrated energy system planning for carbon emission peak and carbon neutrality. *Autom. Electr. Power Syst.* **2022**, *46*, 1–10.
6. Chen, Y.B.; Yao, Y.; Lin, Y.Z.; Yang, X.N. Dynamic state estimation for integrated electricity-gas systems based on Kalman filter. *CSEE J. Power Energy Syst.* **2022**, *8*, 293–303. [CrossRef]
7. Sheng, T.T.; Yin, G.X.; Guo, Q.L.; Sun, H.B.; Pan, Z.G. A hybrid state estimation approach for integrated heat and electricity networks considering time-scale characteristics. *J. Mod. Power Syst. Clean Energy* **2020**, *8*, 636–645. [CrossRef]
8. Dong, J.N.; Sun, H.B.; Guo, Q.L.; Sheng, T.T.; Qiao, Z. State estimation of combined electric-gas networks for energy internet. *Power Syst. Technol.* **2018**, *42*, 400–409. [CrossRef]
9. Dong, J.N.; Sun, H.B.; Guo, Q.L.; Pan, Z.G. State estimation for combined electricity and heat networks. *Power Syst. Technol.* **2016**, *40*, 1635–1641. [CrossRef]
10. Chen, Y.B.; Zheng, S.L.; Yang, N.; Yang, X.N.; Liu, K.C. Robust state estimation of electric-gas integrated energy system based on weighted least absolute value. *Autom. Electr. Power Syst.* **2019**, *43*, 61–70. [CrossRef]
11. Zheng, S.L.; Liu, J.; Chen, Y.B.; Qi, B. Bilinear robust state estimation based on weighted least absolute value for integrated electricity-gas system. *Power Syst. Technol.* **2019**, *43*, 3733–3742. [CrossRef]
12. Du, Y.X.; Zhang, W.; Zhang, T.T. ADMM-based distributed state estimation for integrated energy system. *CSEE J. Power Energy Syst.* **2019**, *5*, 275–283. [CrossRef]
13. Zhang, Y.B. Study on the Methods for Analyzing Combined Gas and Electricity Networks. Ph.D. Thesis, China Electric Power Research Institute, Beijing, China, 2005.
14. Liu, W.J.; Li, P.Y.; Yang, W.T.; Chung, C.Y. Optimal energy flow for integrated energy systems considering gas transients. *IEEE Trans. Power Syst.* **2019**, *34*, 5076–5079. [CrossRef]
15. Ci, W.B.; Gu, H.F.; Zhu, J.S. Multi-timescale state estimation for integrated electricity and heat system. *Therm. Power Gener.* **2021**, *50*, 94–100. [CrossRef]
16. Dong, L.; Wang, C.F.; Li, Y.; Wang, X.Y.; Dong, X.C.; Sun, Y.Y.; Pu, T.J. Multi-snapshot state estimation of electric-gas integrated energy system. *Power Syst. Technol.* **2020**, *44*, 3458–3465. [CrossRef]
17. Yin, G.X.; Chen, B.B.; Sun, H.B.; Guo, Q.L. Energy circuit theory of integrated energy system analysis (IV): Dynamic state estimation of the natural gas network. *Proc. CSEE* **2020**, *40*, 5827–5835. [CrossRef]
18. Liu, X.R.; Li, Y.; Sun, Q.Y.; Pan, Y.L. Interaction and joint state estimation of electric-gas-thermal coupling network. *Power Syst. Technol.* **2021**, *45*, 479–488. [CrossRef]
19. Mike, S.; Paliwal, K.K. Bidirectional recurrent neural networks. *IEEE Trans. Signal Process.* **1997**, *45*, 2673–2681. [CrossRef]
20. Hochreiter, S.; Schmidhuber, J. Long short-term memory. *Neural Comput.* **1997**, *9*, 1735–1780. [CrossRef] [PubMed]
21. Wu, Y.-K.; Wu, Y.-C.; Hong, J.-S.; Phan, L.H.; Quoc, D.P. Probabilistic forecast of wind power generation with data processing and numerical weather predictions. *IEEE Trans. Ind. Appl.* **2021**, *57*, 36–45. [CrossRef]
22. Keddouda, A.; Ihaddadene, R.; Boukhari, A.; Atia, A.; Arıcı, M.; Lebbihiat, N.; Ihaddadene, N. Solar photovoltaic power prediction using artificial neural network and multiple regression considering ambient and operating conditions. *Energy Convers. Manag.* **2023**, *288*, 117186. [CrossRef]
23. Chen, H.; Chang, X. Photovoltaic power prediction of LSTM model based on Pearson feature selection. *Energy Rep.* **2021**, *7*, 1047–1054. [CrossRef]
24. Liu, X.Z.; Wu, J.Z.; Jenkins, N.S.; Bagdanavicius, A. Combined analysis of electricity and heat networks. *Energy Procedia* **2014**, *61*, 155–159. [CrossRef]
25. Wang, D.; Liu, L.; Jia, H.J.; Wang, W.; Zhi, Y.; Meng, Z. Review of key problems related to integrated energy distribution systems. *CSEE J. Power Energy Syst.* **2018**, *4*, 130–145. [CrossRef]

energies

MDPI

Article

Digital Twin-Enabled Distributed Robust Scheduling for Park-Level Integrated Energy Systems

Xiao Chang [1], Shengwen Li [1], Qiang Wang [1], Liang Ji [2,*] and Bitian Huang [2]

[1] State Grid Shanxi Electric Power Company Electric Power Research Institute, Taiyuan 030001, China; x565652025@163.com (X.C.); nan202506@163.com (S.L.); 18135186374@163.com (Q.W.)

[2] Faculty of Electrical Engineering, Shanghai University of Electric Power, Shanghai 200090, China; bitian1205@163.com

* Correspondence: jiliang@shiep.edu.cn

Abstract

With the deepening of multi-energy coupling and the integration of high proportions of renewable energy, the Park Integrated Energy System (PIES) 1demonstrates enhanced energy utilization flexibility. However, the random fluctuations in photovoltaic (PV) output also pose new challenges for system dispatch. Existing distributed robust scheduling approaches largely rely on offline predictive models and therefore lack dynamic correction mechanisms that incorporate real-time operational data. Moreover, the initial probability distribution of PV output is often difficult to obtain accurately, which further degrades scheduling performance. To address these limitations, this paper develops a PV digital twin model capable of providing more accurate and continuously updated initial probability distributions of PV output for distributed robust scheduling in PIESs. Building upon this foundation, this paper proposes a distributed robust scheduling method for the PIES based on digital twins. This approach aims to maximize the flexibility of energy utilization in PIESs and overcome the challenges posed by random fluctuations in PV output to PIES operational scheduling. First, a PIES model is established after investigating a park-level practical integrated energy system. To describe the uncertainty of PV output, a PV digital twin model that incorporates historical data and temporal features is developed. The long short-term memory (LSTM) neural network is employed for output prediction, and real-time data are integrated for dynamic correction. On this basis, error perturbations are introduced, and PV scenario generation and reduction are carried out using Latin hypercube sampling and k-means clustering. To achieve multi-energy cascade utilization, the objective of optimization is defined as the minimization of the sum of system operating cost and curtailment cost. To this end, a two-stage distributed robust optimization model is constructed. The optimal scheduling scheme was obtained by solving the problem using the column-and-constraint generation (CCG) algorithm. The proposed method was finally validated through a case study involving an actual industrial park. The findings indicate that the constructed digital twin model achieves a significant improvement in prediction accuracy compared to traditional models, with the root mean square error and mean absolute error reduced by 13.3% and 10.81%, respectively. Furthermore, the proposed distributed robust scheduling strategy significantly enhances the operational economics of PIESs while maintaining system robustness, compared to conventional methods, thereby demonstrating its practical application value in PIES scheduling.

Keywords: digital twin; park integrated energy systems; output uncertainty; distributed robust optimization; scenario generation

1. Introduction

In the context of global environmental degradation and the pressing need for a transition to low-carbon energy solutions, Park Integrated Energy Systems (PIESs) emerge as a pivotal solution. PIESs are characterized by multi-energy complementarity, cascading energy utilization, enhanced energy efficiency, and increased integration of renewable sources such as photovoltaics and wind power. These features position PIESs as a critical component in achieving carbon peaking and carbon neutrality goals [1,2]. PV generation exhibits significant environmental sensitivity, resulting in inherent fluctuations and randomness. These characteristics pose challenges to the safe and stable operation of power systems. Therefore, incorporating the uncertainty of photovoltaic output into the optimization and scheduling of PIESs is crucial for ensuring the secure and stable operation of the system.

In describing PV output uncertainty, commonly employed methods primarily encompass several major categories including probabilistic, stochastic process-based, and scenario-based approaches. Probabilistic methods, grounded in the statistical characteristics of historical data, fit PV output to specific probability distributions (such as Beta, log-normal, or hybrid distributions), proving suitable for scenarios with ample data and relatively stable distribution characteristics. Stochastic process methods (such as ARIMA, GARCH, Markov chains, Gaussian processes, etc.) can characterize the correlations within time series, but they often prove inadequate for capturing the non-stationarity and extreme volatility inherent in PV output. Scenario-based approaches, generating extensive samples via Monte Carlo simulations, LHS or Copula methods, then extracting a small number of representative scenarios through clustering, currently represent the most widely applied pathway in integrated energy system scheduling. Furthermore, while deep generative models (GANs, VAE) and machine learning methods such as random forests can fit complex distributions, they suffer from high training costs, strong parameter sensitivity, and insufficient interpretability. Consequently, their application in engineering scheduling remains limited.

Compared to traditional Monte Carlo sampling, Latin Hypercube Sampling (LHS) achieves more uniform coverage of values with the same sample size. By partitioning the input variable range into non-overlapping subintervals and sampling within each interval, it ensures balanced distribution of samples, thereby enhancing estimation accuracy and sample utilization efficiency. During the scenario reduction phase, K-means clustering generates a smaller number of well-defined representative scenarios while preserving distributional representativeness. This approach offers good interpretability and significantly reduces the computational scale of the subsequent DRO scheduling model. Compared to complex distribution generation methods such as Copula, GAN, and VAE, K-means offers straightforward computation, intuitive results, and effectively controls the computational scale of the DRO scheduling model, thereby enhancing the convergence efficiency of the CCG algorithm.

In summary, this paper employs LHS for sample generation and combines it with K-means clustering to construct typical photovoltaic scenarios. This approach balances accuracy, interpretability, and algorithmic convergence, rendering it more suitable for the integrated energy system dispatch scenarios described herein.

In PIES optimization scheduling, two main strategies mitigate PV output variability: improving forecasting accuracy and applying advanced optimization techniques. With respect to forecasting, methods can be classified into two categories: physical forecasting and statistical forecasting, based on the differing prediction models [3]. The physical prediction method is an engineering-based approach that utilizes a comprehensive simulation model encompassing the entire process, from solar irradiation to photovoltaic power generation. While it does not require extensive data, it necessitates accurate photovoltaic parameter information and complex photovoltaic mechanism modeling.

Conversely, the statistical prediction method utilizes big data models to ascertain the correlation between photovoltaic power generation inputs and outputs for forecasting. This data-driven approach exhibits a reduced reliance on physics-based modeling. Reference [4] examines the spatiotemporal characteristics of PV output and designs a discriminator structure that integrates CNN and LSTM networks. In this structure, the earth mover's distance is employed as the loss function of the discriminator to generate PV output scenarios. Reference [5] addresses alterations in weather patterns caused by significant meteorological fluctuations by proposing a weather classification algorithm based on multi-scale volatility features. A multi-channel structured LSTM modeling approach is employed for comprehensive PV forecasting. While the aforementioned PV forecasting models take into account a variety of parameters, they are fundamentally reliant on pre-trained models for prediction. Reference [6] employs multiple data-driven models to forecast PV output, including Elastic Net regression, linear regression, random forests, k-nearest neighbors, gradient boosting regressors, lightweight gradient boosting regressors, extreme gradient boosting regressors, and decision tree regressors. This study systematically compares the performance of different machine learning methods in PV forecasting, providing guidance for selecting data-driven predictive models. Reference [7] proposes a novel fractional-order whale optimization algorithm enhanced support vector regression framework. By incorporating fractional calculus into the whale optimization algorithm, it effectively improves the balance between exploration and exploitation during hyperparameter optimization, yielding significantly enhanced prediction accuracy compared to traditional benchmark models.

The aforementioned PV forecasting studies have thoroughly explored aspects such as feature construction, model selection, and parameter optimization, achieving satisfactory prediction outcomes. However, such methods generally rely on static models trained offline, whose predictive performance becomes fixed after model training. When external conditions change—such as abrupt weather shifts, PV module aging, or increased dust contamination—the original model's input-output relationship may become inaccurate. Consequently, prediction accuracy degrades over time, rendering the model incapable of continuously adapting to the dynamic variations inherent in real-world operational scenarios [8].

Therefore, it is necessary to compare the established models with their physical entities in real time, continuously track their accuracy, and perform online corrections when the accuracy requirement is not met. This can be achieved by constructing a digital twin (DT) model of PV, which enables real-time tracking of the physical PV system and iterative correction of the virtual model.

DT principally utilizes historical and real-time data to construct a virtual twin of physical entities, thereby enabling the reflection of their real-time status [9]. Reference [10] provides a comprehensive review of the current application status and key advancements of DT technology within the power generation sector, establishing a novel DT classification framework based on temporal scales and application scenarios. As one of the core enabling technologies for DT, machine learning has demonstrated formidable modeling capabilities across multiple energy contexts. In related research, Reference [11] proposed a novel hybrid quantum genetic algorithm–proximal policy optimization framework. This method combines the efficient global search capability of the quantum genetic algorithm with the adaptive policy update mechanism of proximal policy optimization, thereby achieving automatic parameter tuning and efficient feature space refinement. Reference [12] constructed a high-precision artificial intelligence diagnostic system successfully applied to wind turbine fault detection, demonstrating outstanding intelligent recognition performance. Reference [13] employed advanced deep learning image classification models for screening and evaluation, achieving exceptional results in image processing. These studies have yielded significant outcomes in machine learning algorithm design and model construction. Reference [14] employs advanced analytical modules such as CNN-LSTM and optimization algorithms to construct a proactive cognitive DT system. This system enables forward-looking power grid anomaly identification by predicting future health states and comparing them against real-time monitoring data.

The aforementioned research has achieved significant results in both DT architecture design and intelligent model construction. However, it has primarily focused on application scenarios such as condition monitoring, anomaly detection, or fault identification, without yet integrating DTs with distributed robust scheduling. In other words, how to leverage the real-time calibration capabilities of digital twins to construct uncertainty-driven probabilistic scenarios and serve the distributed robust scheduling of PIESs remains an unaddressed gap in current research.

The PV DT model constructed herein possesses not only self-learning capabilities but also dynamically calibrates itself based on real-time data, thereby ensuring its predictive accuracy remains consistently stable over time. This characteristic provides enhanced adaptability and reliability for PV forecasting. Furthermore, the DT model delivers more precise initial probability information for distributed robust scheduling, thereby improving the economic efficiency and robustness of the scheduling model when confronting PV output uncertainty.

With respect to the optimization techniques employed, the scheduling methods of PIES can be classified according to their uncertainty modeling approaches. These approaches include fuzzy optimization [15], stochastic optimization [16], robust optimization [17], and distributionally robust optimization (DRO) [18]. Fuzzy optimization and stochastic optimization are relatively dependent on subjective factors in the selection of related functions. While these systems offer notable economic advantages, their inherent fragility poses significant challenges to the reliable and secure operation of the system. Conversely, robust optimization addresses the worst-case scenarios, ensuring system resilience but often resulting in overly conservative outcomes and significant economic expenditures. DRO integrates the advantages of both stochastic optimization and robust optimization. The primary focus of this study is the uncertainty surrounding parameters in probability distribution functions. The objective is to ascertain the probability distribution of PV output under the most unfavorable conditions. Reference [19] established a DRO scheduling model for PIESs based on historical wind and solar data. In comparison with alternative optimization methodologies, DRO strikes a balance between economic efficiency and robustness. However, the initial scenarios required for its implementation are often challenging to obtain. Consequently, this paper employs a photovoltaic digital twin model, incorporates error, and utilizes Latin hypercube sampling and k-means clustering to provide typical PV output scenarios and corresponding scenario probabilities for DRO scheduling.

To overcome such problems and challenges, this paper proposes a DRO scheduling approach for PIESs based on DT. First, a DT model of PV is established, and on this basis, initial PV output scenarios are generated to characterize output uncertainty using Latin hypercube sampling and k-means clustering. Then, with the minimization of the total system operating cost and curtailment cost as the optimization objective, a two-stage distributionally robust optimization model is constructed considering the constraints of various equipment in PIES. To demonstrate the effectiveness and engineering applicability of the proposed methodology, a PV DT model and the corresponding DRO scheduling framework were implemented and validated at a representative site within an industrial park in Shanxi Province.

2. Integrated Energy Structure of the Industrial Park

Given that the actual campus energy system underpinning this research does not incorporate combined heat and power (CHP) equipment, this paper builds upon the PIES architecture proposed in reference [20]. It removes CHP-related equipment and their coupling relationships, thereby constructing a system model more suited to the research scenario. The adjusted structure of the PIES is illustrated in Figure 1.

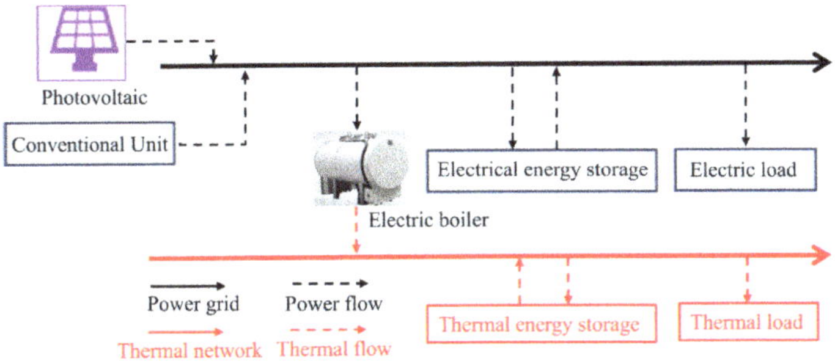

Figure 1. PIES Architecture.

The source side encompasses PV output, conventional power units. The electric boiler functions as a thermoelectric conversion device, thereby facilitating complementary multi-energy flows. Concurrently, thermal and electrical storage devices facilitate the absorption of surplus photovoltaic output, thereby mitigating the fluctuations in photovoltaic power generation.

2.1. Digital Twin Framework of Photovoltaic System

The concept of DTs can be traced back to a proposal by Professor Grieves of the University of Michigan in 2002 regarding product lifecycle management [21,22]. Photovoltaic systems are characterized by their high susceptibility to environmental influences and the presence of complex operational mechanisms. Consequently, it is necessary to extract intrinsic characteristics from the data available in order to establish a data-driven model. A DT primarily achieves this by mining and analyzing vast amounts of real-time and historical data to reflect the real-time status of physical models.

The physical object constructed in this paper is the photovoltaic system within PIES, with the primary focus of this study being the establishment of a digital twin model for its forecasting capabilities. The Digital Twin Framework for PV in PIES is illustrated in Figure 2.

As shown in Figure 2, the DT model constructed herein employs a bidirectional data interaction mechanism: operational data collected in real time at the physical layer drives updates and corrections to the virtual model, whilst predictive outcomes generated at the virtual layer exert a reciprocal influence upon the physical system, guiding its underlying operational strategies. This establishes a closed-loop, collaborative system architecture.

The primary process for constructing a PV DT model is as follows:

First, collect meteorological and historical output data from physical photovoltaic systems, then input this data into the data-driven model shown in Figure 2. This model is based on an LSTM neural network. The advantages of this model are detailed in the next section. After inputting the data, the initial model is trained using the LSTM.

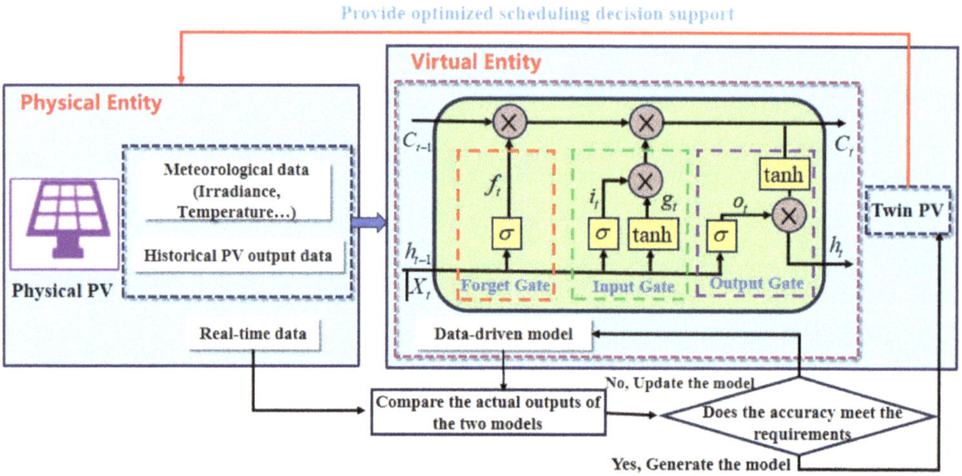

Figure 2. Digital Twin Framework for PV in PIES.

Employ the following formula for error assessment and error classification.

$$e(t) = \left| P_i^{pv} - \hat{P}_i^{pv} \right| \quad (1)$$

$$e(t) > \varepsilon \quad (2)$$

where P_i^{pv} denotes the actual PV output, $\hat{P}_i^{pv}$ represents the predicted PV output, ε denotes the error threshold.

An error assessment is conducted every five minutes. If the error meets the required level of precision, the final DT model is generated. Otherwise, the data-driven model is updated by adjusting its hyperparameters through Bayesian optimization and retraining. If the error consistently fails to meet the requirements, comprehensive retraining is performed at hourly intervals. It is important to note that this data-driven model is not static but rather undergoes real-time updates.

To evaluate the performance of the DT model constructed in this paper for PV output forecasting, the root mean square error (RMSE) and mean absolute error (MAE) were employed as evaluation metrics to quantitatively assess its predictive accuracy, specifically, as shown in the following equation:

$$RMSE = \sqrt{\frac{1}{n}\sum_{i=1}^{n} (P_i^{pv} - \hat{P}_i^{pv})^2} \quad (3)$$

$$MAE = \frac{1}{n}\sum_{i=1}^{n} \left| P_i^{pv} - \hat{P}_i^{pv} \right| \quad (4)$$

2.2. Photovoltaic Power Forecasting Based on LSTM

The photovoltaic power output demonstrates significant diurnal periodicity and seasonal variations and is also influenced by short-term meteorological changes. In order to circumvent issues such as gradient vanishing and gradient explosion that commonly occur during the training process of traditional neural networks, a memory unit and gating mechanism were introduced, leading to the development of LSTM. LSTM has been shown to enhance stability and applicability when processing time series data by forgetting unnecessary information while retaining key features. The LSTM architecture is illustrated in Figure 3.

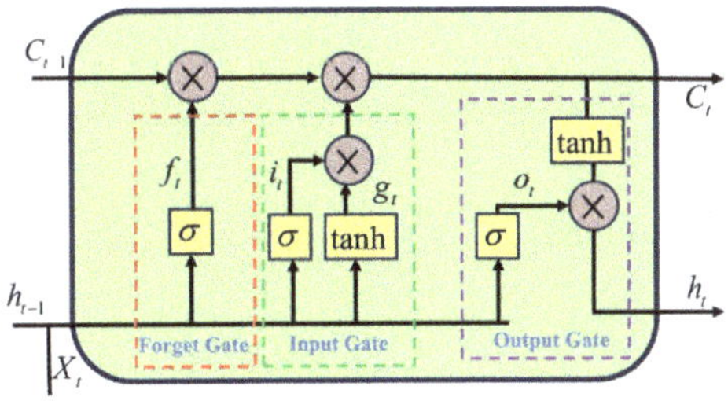

Figure 3. LSTM architecture.

In Figure 3, C_t, h_t, and X_t represent the cell state, output, and input at time t. f_t, i_t, g_t, and o_t denote the computational formulas for each gate unit. The specific formulas are provided in References [23,24].

2.3. Scenario Generation Based on Latin Hypercube Sampling and Scenario Reduction with K-Means Clustering

In order to account for subsequent scheduling robustness requirements, it is necessary to superimpose error margins on the PV forecast. However, PV forecast results incorporating uncertainty errors are calculated during the scheduling process. Consequently, multi-scenario techniques can be utilized to transform PV forecasts with uncertainty errors into a set of deterministic scenarios [25,26]. The prediction error of PV output can be modeled as a population following a normal distribution with zero mean. Consequently, the actual PV power can be expressed as in Equation (5).

$$P_{\mathrm{pv},t} = \hat{P}_{\mathrm{pv},t} + \varepsilon_t, \varepsilon_t \sim N(0, \sigma^2) \tag{5}$$

where $P_{\mathrm{pv},t}$ represents the actual PV output, $\hat{P}_{\mathrm{pv},t}$ represents the predicted PV output, ε_t is the random disturbance at time t and σ is the standard deviation in the normal distribution.

The Latin hypercube sampling (LHS) method can be utilized to generate PV output scenarios that align with the distribution of PV forecasting errors. In comparison with Monte Carlo simulation sampling, LHS attains a more uniform distribution of values while maintaining an equivalent sample size. The range of each input variable is divided into multiple non-overlapping subintervals, and a single sample is randomly selected from each subinterval. This stratified sampling technique ensures that all sampling regions are covered by the sampling points, thereby improving estimation accuracy and sample utilization.

The scenarios generated by LHS are frequently characterized by their expansive scale. The utilization of raw scenarios invariably leads to a marked reduction in the efficiency and speed of solutions. Therefore, it is necessary to reduce the large number of original scenarios by selecting a few representative scenarios and assigning corresponding probability values to each. In this paper, the k-means clustering algorithm is employed for scenario reduction. The k-means clustering algorithm first determines the initial cluster centers using the elbow method then computes the Euclidean distance between the remaining scenarios and the cluster centers, assigning each scenario to the nearest cluster center. The specific implementation of this method is described in reference [27].

The contour coefficient serves to describe the clarity of each cluster's outline following clustering, constituting a vital metric for evaluating clustering efficacy. Its core function lies in simultaneously measuring intra-cluster compactness and inter-cluster separation, thereby determining the validity of the clustering. The contour coefficient $S(i)$ comprises two components: the intraclass average distance $a(i)$ and the interclass minimum distance $b(i)$. The specific expression is as follows:

$$a(i) = \frac{1}{|C_i| - 1} \sum_{j \in C_i, j \neq i} d(i,j) \tag{6}$$

$$b(i) = \min_{k \neq C_i} \left(\frac{1}{|C_k|}\right) \sum_{j \in C_k} d(i,j) \tag{7}$$

$$S(i) = \frac{b(i) - a(i)}{\max(b(i), a(i))} \tag{8}$$

where $d(i,j)$ denotes the distance from i to j, while C_i represents the cluster to which sample i belongs. A smaller value of a indicates $a(i)$ more compact cluster. $b(i)$ denotes the average distance between sample i and all samples in other clusters, taking the minimum value. A larger value of $b(i)$ indicates more pronounced separation from other clusters. The larger the $S(i)$ value defined by combining these two parameters, the more reasonable the clustering result for that sample and the better the overall clustering performance.

3. Distributed Robust Scheduling Model for Integrated Energy Systems in Industrial Parks

The overall logic of the distributed robust scheduling model for PIES is demonstrated in Figure 4. Firstly, a digital twin model for photovoltaic generation is established, along with a basic PIES scheduling model that aims to minimize system operational costs without considering the uncertainty in photovoltaic output. As a consequence of photovoltaic power forecasting, errors are introduced, and multiple scenarios are generated to match the error distribution. These scenarios are then reduced to a small number of representative ones, each assigned a corresponding probability. The inherent variability in photovoltaic output is then integrated into the fundamental PIES scheduling model, thereby transforming it into a distributed, robust scheduling model for PIES. Ultimately, the CCG algorithm is employed to solve and obtain the final distributed robust scheduling solution for PIES.

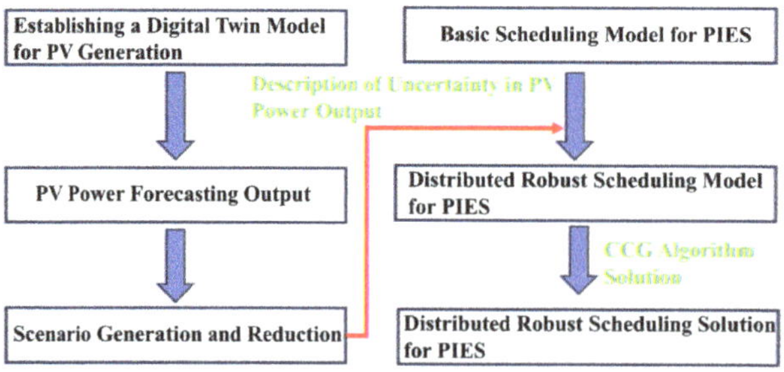

Figure 4. Overall Logic of PIES' Distributed Robust Scheduling.

3.1. *Dispatch Model for Integrated Energy Systems in Industrial Parks*

(1) Objective function

This section first constructs the basic scheduling model for PIES without considering the uncertainty in PV power output, which serves as the foundation for the subsequent distributed scheduling. The objective function of this model is to minimize system operational costs and curtailment costs, as shown in the following equation.

$$F_{\min} = F_{\text{op}} + F_{\text{qv}} \tag{9}$$

$$F_{\text{op}} = F_{\text{qt}}^{\text{g}} + F_{\text{g}}^{\text{g}} \tag{10}$$

where F_{op} represents the system operating cost, and F_{qv} represents the curtailment cost. The system operating cost comprises the start-up and shutdown costs of conventional generating units, and the power generation cost, denoted by F_{qt}^{g} and, respectively.

The generation cost can be represented by the corresponding fuel coefficient.

The generation costs for conventional units are expressed as follows:

$$F_{\text{g}}^{\text{g}} = \sum_{t=1}^{T} \sum_{j}^{N_{\text{g}}} (a_j^{\text{g}} (P_{j,t}^{\text{g}})^2 + b_j^{\text{g}} P_{j,t}^{\text{g}} + C_{j,t}^{\text{g}}) \tag{11}$$

where a_j^{g} and b_j^{g} represent the equivalent nonlinear and linear power generation cost functions for the j-th conventional unit, respectively. $C_{j,t}^{g}$ denote fixed costs for the j-th unit generators such as maintenance expenses. T denotes the total number of time periods. N_{g} denote the total number of units for conventional. $P_{j,t}^{\text{g}}$ denote the power output of the j-th unit for conventional units at time t.

In order to address potential fluctuations, conventional units incur corresponding startup and shutdown costs.

The startup and shutdown costs are given as

$$F_{qt}^{g} = \sum_{t=1}^{T} \sum_{j}^{N_{\text{g}}} (R_{\text{ks}} s_{j,t}^{\text{on}} + R_{\text{ts}} s_{j,t}^{\text{off}}) \tag{12}$$

where R_{ks} and R_{ts} represent the startup and shutdown costs for conventional units, while $s_{j,t}^{on}$ and $s_{j,t}^{off}$ denote the startup/shutdown status flags of the j-th unit for conventional units at time t. When $s_{j,t}^{\text{on}}$ equals 1 and $s_{j,t}^{\text{off}}$ equals 0, it indicates startup; when $s_{j,t}^{\text{off}}$ equals 1 and $s_{j,t}^{\text{on}}$ equals 0, it indicates shutdown.

In order to achieve greater decarbonization, it is essential to prioritize the maximization of PV power accommodation. Consequently, the financial implications of PV curtailment are integrated into the objective function.

$$F_{qv} = \sum_{t=1}^{T} \sum_{j}^{N_{\text{pv}}} \beta (P_{j,t}^{\text{pv}} - \hat{P}_{j,t}^{\text{pv}}) \tag{13}$$

where N_{pv} denotes the number of PV units, β represents the linear coefficient of curtailment costs, $P_{j,t}^{pv}$ and $P_{j,t}^{pv}$ denote the actual output and predicted output of the j-th PV unit at time t, respectively.

(2) Constraints

To ensure the safe and stable operation of the system and to account for the inherent characteristics of various devices, the PIES scheduling must satisfy a set of constraint conditions. The power balance constraints include both electrical power balance and thermal power balance.

The electrical power balance constraint is given as

$$\sum_{j}^{N_g} P_{j,t}^{g} + \sum_{j}^{N_{pv}} P_{j,t}^{pv} + \sum_{j}^{N_{es}} s_{j,t}^{cs} P_{j,t}^{es} - \sum_{j}^{N_{eb}} P_{j,t}^{eb} = P_t^{l} \tag{14}$$

where N_{es} and N_{eb} represent the number of electrical energy storage units and electric boilers, respectively. $s_{j,t}^{cs}$ denotes the j-th energy storage charge/discharge indicator at time, where -1 indicates charging and +1 indicates discharging. $P_{j,t}^{es}$ is the output value of the j-th electrical energy storage unit at time t. $P_{j,t}^{eb}$ is the electrical power consumed by the j-th electric boiler at time t. P_t^{l} is the electrical load at time t.

The thermal power balance constraint is given as

$$\sum_{j}^{Nes} s_{j,t}^{cs} W_{j,t}^{es} + \sum_{j}^{Neb} W_{j,t}^{eb} = W_t^{l} \tag{15}$$

where $W_{j,t}^{es}$ denotes the thermal output of the j-th thermal energy storage unit at time t; $W_{j,t}^{eb}$ is the thermal output of the j-th electric boiler at time t; and W_t^{l} represents the thermal load at time t.

Conventional units must satisfy corresponding electrical power constraints and ramping constraints.

Electrical power constraints are defined as:

$$P_{j,\min}^{g} \le P_{j,t}^{g} \le P_{j,\max}^{g} \tag{16}$$

where $P_{j,\min}^{g}$ and $P_{j,\max}^{g}$ represent the upper and lower limits of the output of the j-th unit, respectively.

Ramp rate constraints:

$$Z_{j,h}^{g} \le P_{j,t}^{g} - P_{j,t-1}^{g} \le Z_{j,p}^{g} \tag{17}$$

where $Z_{j,p}^{g}$ denotes the climbing rate of a conventional unit, and $Z_{j,h}^{g}$ denotes the sliding rate.

The start-up and shutdown status of conventional units is represented by a single state variable, namely:

$$R_{ks} + R_{ts} \le 1 \tag{18}$$

The output of the electric boiler must also meet the upper and lower limit requirements and coupling requirements, with its coupling efficiency also being a fixed value:

$$\begin{cases} P_{j,\min}^{eb} \le P_{j,t}^{eb} \le P_{j,\max}^{eb} \\ W_{j,t}^{eb} = \eta^{eb} P_{j,t}^{eb} \end{cases} \tag{19}$$

where $P_{j,\min}^{eb}$ and $P_{j,\max}^{eb}$ represent the upper and lower limits of the output for the j-th electric boiler unit, respectively, and η^{eb} denotes the electrical-to-thermal conversion efficiency of the electric boiler.

Photovoltaic output must meet its upper and lower limit requirements:

$$P_{j,\min}^{pv} \le P_{j,t}^{pv} \le P_{j,\max}^{pv} \tag{20}$$

where $P_{j,\min}^{pv}$ and $P_{j,\max}^{pv}$ represent the upper and lower limits of the j-th photovoltaic output, respectively.

The operation of energy storage systems must satisfy the mutually exclusive charging and discharging requirement, as well as the upper and lower limits of charging/discharging power and storage capacity. In addition, the initial and final states of the storage system must be consistent. Since the operational principles of electrical and thermal energy storage are similar, this paper takes electrical energy storage as an example, and the thermal energy storage model is therefore not presented in detail.

The electrical energy storage constraints as follows:

$$\begin{cases} s^{ch} + s^{dis} \leq 1 \\ s^{cs} = s^{ch} - s^{dis} \\ s^{ch} * P^{ch}_{j,\min} \leq s^{ch} * P^{ch}_{j,t} \leq s^{ch} * P^{ch}_{j,\max} \\ s^{dis} * P^{dis}_{j,\min} \leq s^{dis} * P^{dis}_{j,t} \leq s^{ch} * P^{dis}_{j,\max} \\ R_{Lj,t+1} = R_{Lj,t} + \eta^{ch} P^{ch} - \eta^{dis} P^{dis} \\ R_{Lj,\min} \leq R_{Lj,t} \leq R_{Lj,\max} \\ R_{Lj,t=0} = R_{Lj,t=T} \end{cases} \tag{21}$$

where s^{ch}/s^{dis} are the energy storage charge/discharge status bits. A value of 1 indicates charging/discharging, while a value of 0 indicates no charging/discharging. $P^{ch}_{j,t}$ and $P^{dis}_{j,\max}$ denote the charging/discharging power of the j-th energy storage unit at time t. $P^{ch}_{j,\min}$, $P^{ch}_{j,\max}$, $P^{dis}_{j,\min}$, and $P^{dis}_{j,\max}$ represent the upper and lower limits of charging/discharging for the j-th unit. η^{ch} and η^{dis} are the charging/discharging conversion coefficients. $R_{Lj,t}$ indicates the capacity of the j-th unit at time t. $R_{Lj,t=0}$ and $R_{Lj,t=T}$ denote the initial and final storage capacities of the j-th unit.

To simplify the analysis, the basic scheduling model underlying the above construction can be expressed by the following equation:

$$\min_{x,y}(a^T x + b^T y) \tag{22}$$

$$Ux \leq 1 \tag{23}$$

$$By \leq C\xi \tag{24}$$

$$Vx + Gy \leq h \tag{25}$$

$$Mx + Ly = l \tag{26}$$

where x represents deterministic variables that are not affected by PV forecast errors, including the start-up and shutdown status of conventional units as well as the charging and discharging states of electrical and thermal energy storage systems. y represents real-time adjustable variables that vary with PV output, including the power outputs of various units. a^T denotes the equivalent start-up and shutdown costs coefficient, while b^T represents the generation cost and the cost associated with PV curtailment. ξ is the PV forecast vector. B, C, V, G, M, and L are coefficient matrices.

Equation (22) represents the objective function, while Equations (23)–(26) correspond to the constraint conditions. Equation (23) describes the start-up/shutdown and charging/discharging state constraints, corresponding to Equations (18) and (21). Equation (24) represents the upper limit constraint of PV output, corresponding to Equation (20). Equations (25) and (26) encompass all the remaining equality and inequality constraints that must be satisfied, including power balance constraints.

3.2. Distributed Robust Dispatch Model for Integrated Energy Systems in Industrial Parks

Even after prediction using DT, PV output still contains uncertainties. Therefore, a distributionally robust approach is adopted to account for this portion of uncertainty. In the previous section, a basic scheduling model was established; based on this model, a distributionally robust model will be constructed.

Distributed robust models combine robust optimization methods with stochastic optimization methods, deriving scheduling schemes by seeking the most unfavorable probability distribution of PV output.

To facilitate analysis and solution, this paper constructs a two-stage DRO scheduling framework: in the first stage, decision variable x remains constant regardless of actual scenario variations; in the second stage, decision variable y is adjusted according to different scenarios to describe feasible scheduling strategies under the worst-case probability distribution. This model identifies optimal scheduling solutions that balance economic efficiency and robustness by searching for the probability distribution of the worst-case scenario within the uncertainty set.

Building upon this foundation, Equation (22) can be further modified as follows:

$$\min_{x\in X}\left[a^T x + \max_{\{p_k\in\Omega\}}\sum_{k=1}^{K} p_k \min_{y_k\in Y}(b^T y_k)\right] \tag{27}$$

where X and Y denote the feasible domains of x and y_k formed by their respective constraints; p_k represents the probability of the scenario occurring; Ω denotes the set of feasible domains for the scenario's probability distribution.

Equation (27) represents the two-stage distributed robust model constructed in this paper. In the first stage, x remains constant regardless of actual scenario changes, while y represents the quantity subject to change in the second stage. This primarily involves probabilistically modeling the worst-case scenario to identify an optimized value that meets economic objectives.

To identify the worst-case probability distribution of the scenarios, the most straightforward method is to enumerate the probability values of each scenario and iterate. However, to ensure the reasonableness of the probability distribution, two additional constraints—the $1-$ norm and $\infty-$ norm—are imposed on p_k based on its initial value. The initial value p_{k0} is obtained from the PV output predicted by the digital twin in Section 2, after clustering and reduction.

To avoid the occurrence of extreme conditions, both the $1-$ norm and the $\infty-$ norm must be considered simultaneously. Their constraints can be expressed as

$$\Omega = \left\{ \{p_k\} \left| \begin{array}{l} p_k \geq 0, k = 1,2,\ldots,K \\ \sum\limits_{k=1}^{K} p_k = 1 \\ \sum\limits_{k=1}^{K} |p_k - p_k^0| \leq \theta_1 \\ \max\limits_{1\leq k\leq K} |p_k - p_k^0| \leq \theta_\infty \end{array} \right. \right\} \tag{28}$$

where θ_1 and θ_∞ represent the probability-allowed deviation values under the $1-$ norm and $\infty-$ norm conditions, respectively.

According to Reference [28], the confidence constraints of θ_1 and θ_∞ satisfy

$$\Pr\left\{\sum_{k=1}^{K}\left|p_k - p_k^0\right| \leq \theta_1\right\} \geq 1 - 2K\exp(-2N\theta_1/K) \tag{29}$$

$$\Pr\left\{\max_{1\le k\le K}\left|p_k - p_k^0\right| \le \theta_\infty\right\} \ge 1 - 2K\exp(-2N\theta_\infty) \tag{30}$$

where Pr is the opportunity constraint function.

The right-hand side of Equations (29) and (30) can also be expressed in terms of the confidence levels α_1 and α_∞ of the uncertainties.

Consequently, θ_1 and θ_∞ can be determined as

$$\begin{cases} \theta_1 = (K/2N)\ln\frac{2K}{1-\alpha_1} \\ \theta_\infty = (1/2N)\ln\frac{2K}{1-\alpha_\infty} \end{cases} \tag{31}$$

It is worth noting that Equations (29) and (30) also contain absolute value variables, which complicate the solution process. Therefore, two auxiliary variables v_k^+ and v_k^- are introduced into these constraints. This allows the absolute values to be transformed into

$$\begin{cases} v_k^+ + v_k^- \le 1 \\ p_k = p_k^0 + p_k^+ - p_k^- \\ 0 \le p_k^+ \le v_k^+\theta_1 \\ 0 \le p_k^- \le v_k^-\theta_1 \end{cases} \tag{32}$$

where p_k^+ and p_k^- denote the positive and negative offsets of the initial values.

Similarly, it can be obtained that:

$$\begin{cases} n_k^+ + n_k^- \le 1 \\ p_k = p_k^0 + p_k^+ - p_k^- \\ 0 \le p_k^+ \le n_k^+\theta_\infty \\ 0 \le p_k^- \le n_k^-\theta_\infty \end{cases} \tag{33}$$

where n_k^+ and n_k^- are also auxiliary variables.

3.3. Model Solution

Equation (27) can be considered as a min-max-min problem. For this problem structure, the column-and-constraint generation (CCG) algorithm is highly suitable. This is due to the fact that it facilitates the decomposition of the original problem into a primary problem and a series of subsidiary problems. The approach entails the alternating resolution of the primary and secondary problems, thereby circumventing the computational explosion that would otherwise ensue from the direct solution of a large-scale model. This, in turn, enhances solution efficiency. The solution process to this problem is outlined below:

(1) Main Problem (MP)

MP lies in obtaining variables that satisfy optimal economic conditions under the most adverse probability distributions. It provides a lower bound (LB) for model Equation (27), as shown in Equation (34).

$$\begin{cases} \min\limits_{x\in X, y_k\in Y}\left[a^T x + SP\right] \\ SP \ge \sum\limits_{k=1}^{K} p_k^c \min\limits_{y_k\in Y}(b^T y_k^c) \end{cases} \tag{34}$$

where c is the number of iterations.

(2) Subproblem (SP)

The subproblem essentially involves identifying the most unfavorable probability distribution while fixing the variable x, thereby providing an upper bound (UB) for Equation (27) to facilitate further iterative computation in the main problem. Specifically, as demonstrated in Equation (35):

$$SP(x^*) = \max_{\{p_k \in \Omega\}} \sum_{k=1}^{K} p_k \min_{y_k \in Y(x^*)} (b^T y_k) \tag{35}$$

First set $LB = 0$ and $UB = \infty$. Then fix SP to solve the main problem Equation (27) to obtain x^*. Update the value of LB such that $LB = \max(P_{\min}, a^T x^* + LP)$. At this point, fix the obtained x^* and solve the subproblem Equation (28) to obtain a new $SP(x^*)$ value. Update the value of UB such that $UB = \min(P_{\max}, a^T x^* + LP(x^*))$.

(3) After obtaining the new $SP(x^*)$ value, solve Equation (35) to obtain the new value, and update the values of LB and UB. Repeat this process until $UB - LB \leq \varepsilon$.

It is posited that, by means of this iterative, alternating solution process, an optimal scheduling scheme that satisfies economic efficiency can ultimately be obtained.

4. Case Study

4.1. Scenario Description

This case study selects an industrial park in Shanxi as the research object. The system includes photovoltaic units, electric boilers, and electrical and thermal energy storage systems.

The operational parameters of these devices are shown in Tables 1–3.

Table 1. Parameters of Conventional Units.

Maximum Electrical Power/*kw*	Minimum Electrical Power/*kw*	a_j^g	b_j^g	$C_{j,t}^g$	Start-Up Cost/CNY	Ramp Rate
1000	30	0.002	0.01	0.2	60	70

Table 2. Energy Storage Parameters.

Device	Maximum Charge/Discharge Power/*kw*	Maximum Capacity/*kw*	Minimum Capacity/*kw*	Charging Efficiency	Discharging Efficiency
Electrical Energy Storage	300	900	50	0.95	0.95
Thermal Energy Storage	200	800	30	0.85	0.9

Table 3. Electric Boiler Operating Parameters.

Maximum Electrical Power/*kw*	Minimum Electrical Power/*kw*	Conversion Efficiency
400	0	0.9

The load curve for the example is shown below Figure 5:

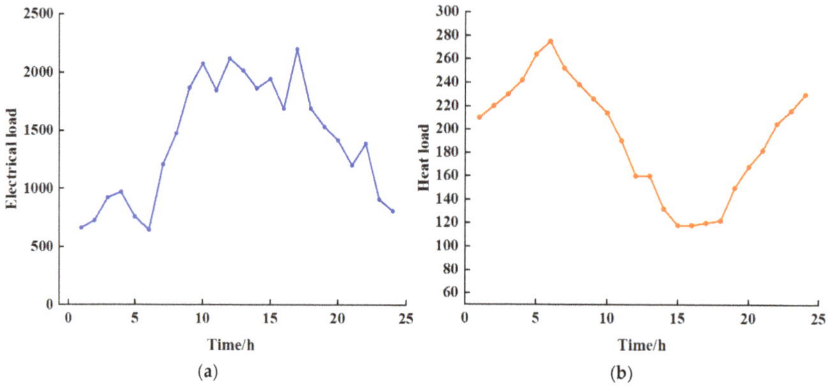

Figure 5. (**a**) Electricity load curve; (**b**) Heat load curve.

4.2. Results of PV Output Scenario Construction

Using solar irradiance, temperature, and humidity as input variables and PV output power as the output variable, an LSTM model is constructed. The data resolution is 5 min, with 70% of the data used as the training set and the remaining 30% as the test set. The time window is set to three steps (equivalent to 15 min) using a single-step forecasting approach. To prevent overfitting, an early stopping mechanism is incorporated into the training of the model. This mechanism monitors the loss in the validation set after each epoch; if the loss in the validation set fails to improve over five consecutive epochs (patience = 5), training is terminated prematurely. All input features undergo standardization via the maximum–minimum normalization method, expressed in Equation (36).

$$x' = \frac{x - x_{\text{min}}}{x_{\text{max}} - x_{\text{min}}} \tag{36}$$

During model training, Adam was used as the optimizer and MAE was used as the loss function. Both MAE and RMSE were used as evaluation metrics. To validate the advantages of the proposed DT model's predictive performance over traditional LSTMs, comparative experiments were conducted using a baseline LSTM model and the prediction model within the digital twin framework. Both models were configured with 200 hidden units and a maximum iteration count of 64. The comparison chart of DT and traditional LSTM predictions is presented in Figure 6.

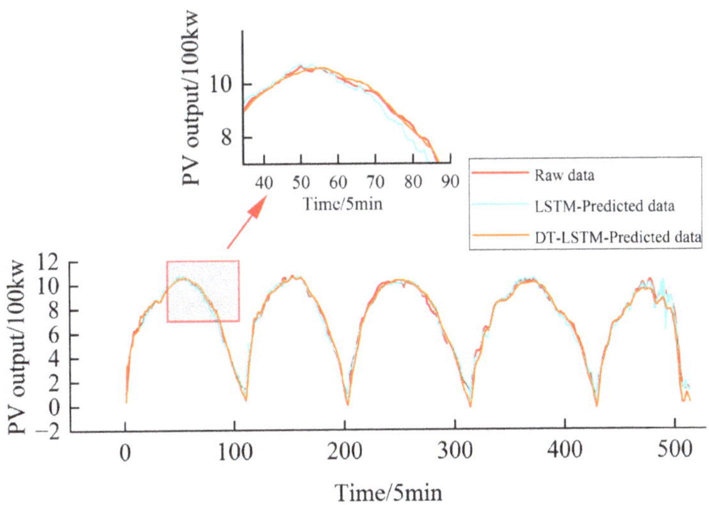

Figure 6. The comparison chart of DT and traditional LSTM predictions.

The comparisons of prediction error results are listed in Table 4. This enables a straightforward comparison of the predictive accuracy differences between various models.

Table 4. Comparison of Prediction Error Results.

Methods	RMSE	MAE
LSTM	0.5837	0.4057
DT-LSTM	0.5837	0.3661

As shown in Table 4, the DT model demonstrates a marked improvement in prediction error compared to the baseline LSTM, with RMSE and MAE reduced by 13.3% and 10.81%, respectively. This outcome indicates that the DT framework enhances both the robustness and predictive accuracy of the model.

Based on the PV output forecast results, LHS was employed to generate 1000 PV output scenarios conforming to the PV forecast output error distribution. With K values set to equal intervals from 2 to 7, elbow diagrams based on the sum of squared errors and contour coefficient diagrams were plotted, as shown in Figures 7 and 8.

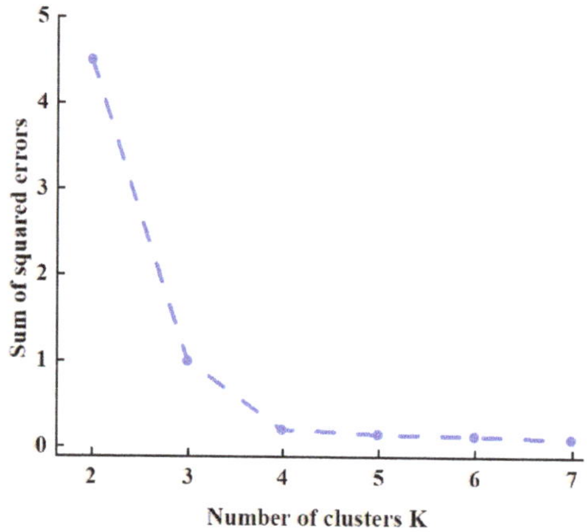

Figure 7. Elbow diagrams based on the sum of squared errors.

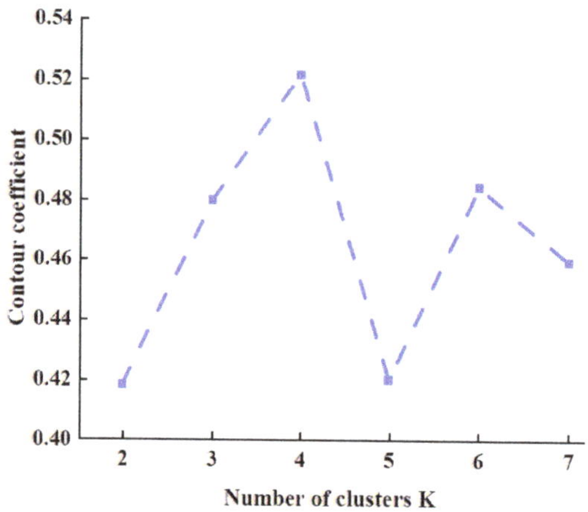

Figure 8. Contour coefficient diagram.

The results of the elbow method shown in Figure 6 reveal that the sum of squared errors exhibits the most pronounced inflection point at K = 4, indicating that the marginal benefit of further increasing the number of clusters for improving clustering performance has markedly diminished. Concurrently, the contour coefficient evaluation in Figure 7 reveals that the average contour coefficient attains its maximum value when K = 4, indicating this clustering number achieves optimal intra-cluster compactness and inter-cluster separation. Combining these two metrics, K = 4 is ultimately selected as the optimal scenario count. Following determination of the cluster number, the k-means method is applied to cluster the PV output data, yielding the initial PV scenarios depicted in Figure 9.

The employment of scenario reduction techniques facilitates the acquisition of the initial PV output scenarios and their probability distributions. This provides the initial probability values and distribution scenarios for distributed robust PV scheduling, facilitating the subsequent derivation of a distributed robust scheduling solution.

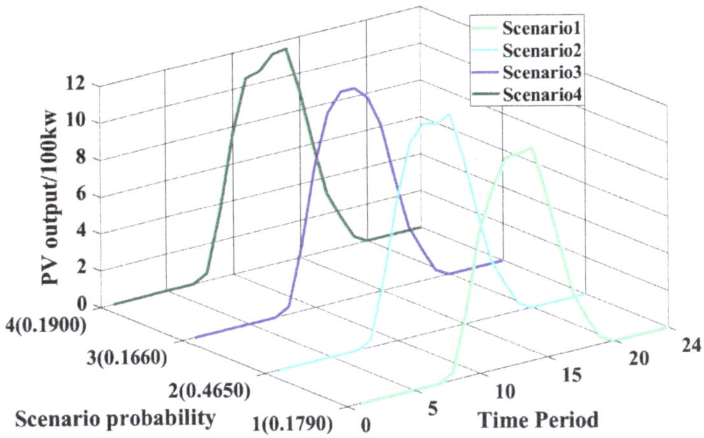

Figure 9. Initial Scenario of PV Output.

4.3. *Analysis of Distributed Robust Scheduling Performance*

Following the acquisition of the initial photovoltaic output scenario and initial probabilities in the preceding section, distributed robust scheduling was performed on the integrated energy system.

The integrated energy power optimization results are demonstrated in Figure 10:

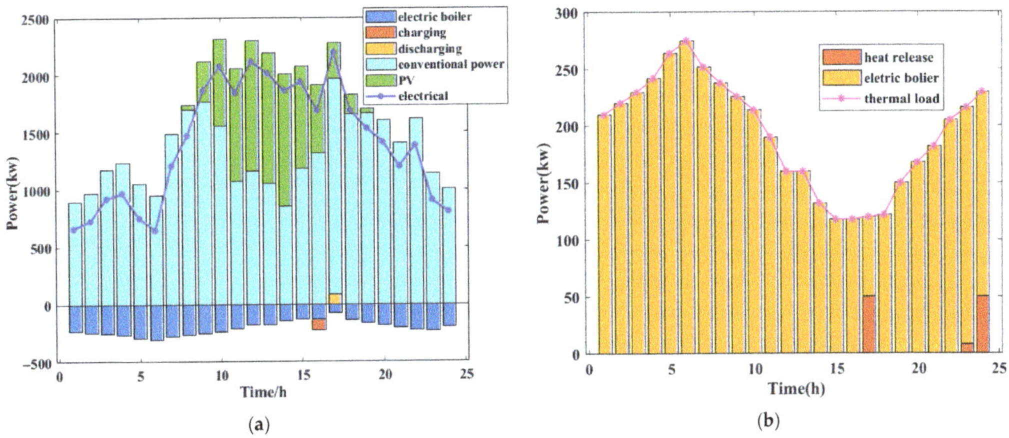

Figure 10. (**a**) Electrical power optimization results; (**b**) Thermal power optimization results.

From the electrical power distribution results in Figure 10, it can be observed that the system consistently adheres to an operational strategy of 'prioritizing the consumption of renewable energy' on the power supply side. Due to the existence of curtailment costs, when PV output is high, the PV electricity first satisfies the system's load demand, reflecting the priority of renewable energy in dispatch. For portions of PV output exceeding real-time demand, the dispatch strategy proactively activates the electrical energy storage system for charging, absorbing and storing surplus electricity. This approach not only prevents curtailment but also enhances local consumption capacity for new energy sources. During periods of low or fluctuating PV output, conventional generating units assume primary responsibility for power supply, effectively compensating for shortfalls caused by renewable energy fluctuations. This ensures the stability of power system operation and maintains supply reliability.

In terms of thermal power scheduling, Figure 10 illustrates that thermal loads are primarily supplied by electric boilers. These boilers convert electrical power into thermal energy, serving as the key equipment for electricity-heat coupling. When thermal load levels within the system are high and electrical supply cannot fully meet thermal demand, thermal storage units release previously stored heat to bridge the thermal gap, thereby maintaining thermal equilibrium. During periods of high PV output, low electricity prices, or energy storage charging, electric boilers can utilize inexpensive or surplus electricity for heating and store it in thermal storage units, thereby achieving time-shifted energy utilization.

The synergy between electricity and heat enhances the flexibility of energy systems. The coordinated operation of electrical energy storage and thermal storage devices not only smooths out the random fluctuations in photovoltaic output but also deepens the coupling between electrical and thermal systems. This improves the efficiency of cascading energy utilization and significantly reduces energy wastage such as curtailed solar power and unused thermal energy. Overall, the dispatch strategy achieves an effective trade-off between renewable energy integration, electricity-heat balance, and operational economics, demonstrating the advantages of coordinated optimization within integrated energy systems.

4.4. Comparative Analysis with Other Methods

To validate the effectiveness of the proposed method, it is compared with both fuzzy optimization and robust optimization approaches. The comparison results are shown in Table 5:

Table 5. Optimization Costs of Different Optimization Methods.

Optimization Methods/Yuan	Optimization Costs/Yuan
Fuzzy Optimization	48,307
Distributed Optimization	53,520
Robust Optimization	56,260

Table 5 demonstrates that the three methods exhibit distinct differences in scheduling costs: fuzzy optimization incurs the lowest cost, robust optimization the highest, while distributed robust optimization's cost lies between the two. This outcome is closely linked to the distinct approaches each method employs in handling PV output uncertainty.

Firstly, fuzzy optimization empirically compresses PV output uncertainty into a single fuzzy value or membership interval, inherently disregarding extreme or worst-case output scenarios. Consequently, the optimization process adopts relatively aggressive scheduling strategies, minimizing costs associated with unit output and energy storage regulation. However, fuzzy optimization offers limited coverage of uncertainty. Should actual PV output deviate from projections, the system becomes more susceptible to power shortfalls or insufficient reserves, exhibiting weaker robustness and heightened operational risks.

Traditional robust optimization usually uses box-shaped uncertainty sets, considering all potential PV errors to be worst-case scenarios. This 'global worst-case' assumption significantly increases reserve requirements and often results in unit dispatch based on extreme conditions. Although it offers high safety margins, the drawbacks are equally pronounced:

dispatch plans become excessively conservative, substantially increasing economic costs—a phenomenon that is particularly evident in systems with high PV penetration.

In contrast, DRO improves robustness by introducing scenario probability distributions and their uncertainty sets. It focuses solely on 'worst-case scenarios with a certain probability'. This approach avoids the excessive conservatism of traditional robust optimization and outperforms the lack of robustness of fuzzy optimization. Within DRO, modifying robust boundaries allows models to reduce unnecessary standby costs while maintaining reliability, thus striking a balance between economic efficiency and robustness.

In summary, DRO effectively mitigates the risk of excessive conservatism in scheduling schemes while ensuring system safety. Of the three approaches, it is the optimal strategy, balancing economic efficiency, flexibility and robustness. Clearly, the DRO method successfully reconciles economic considerations with robustness.

4.5. Analysis of Solution Algorithms

An analysis of the CCG algorithm reveals that the specific results of the iterative cases described herein are presented in Table 6.

Table 6. CCG Algorithm Iteration Results.

Number of Iterations	UB Cost/Yuan	LB Cost/Yuan
1	54,098	26,780
2	53,872	53,380
3	53,520	53,520

As shown in Table 6, after three iterations, the difference between the upper and lower bounds falls below the convergence criterion. The termination of iterations indicates that the CCG algorithm achieves rapid solution speeds for the DRO problem addressed in this paper.

5. Conclusions

This paper develops a digital twin model for PV systems, which, in contrast to traditional models, allows for adjustments over time. Based on forecasting, the Latin Hypercube Sampling and k-means clustering methods are employed to generate initial scenarios and probabilities for PV output.

A comprehensive energy system for the park has been established, and the uncertainty of PV output has been addressed using a distributed robust optimization method with norm-based constraints. This has resulted in a distributed, robust scheduling solution that has been solved using the CCG algorithm.

Through comparative analysis, it is concluded that the distributed robust approach, compared to traditional methods, better balances both the economic efficiency and robustness of the system.

Author Contributions: Conceptualization, S.L.; Methodology, X.C. and Q.W.; Investigation, L.J.; Writing—review and editing, B.H. All authors have read and agreed to the published version of the manuscript.

Funding: This work is supported by State Grid Shanxi Electric Power Company Science and Technology Project Research (52053024000V).

Data Availability Statement: The original contributions presented in this study are included in the article. Further inquiries can be directed to the corresponding author.

Acknowledgments: This work is supported by State Grid Shanxi Electric Power Company Science and Technology Project Research (52053024000V). We would like to thank the State Grid Shanxi Electric Power Company Electric Power Research Institute and Shanghai University of Electric Power for their assistance.

Conflicts of Interest: Authors Xiao Chang, Shengwen Li and Qiang Wang were employed by the State Grid Shanxi Electric Power Company. The remaining authors declare that the research was conducted in the absence of any commercial or financial relationships that could be construed as a potential conflict of interest.

Abbreviations

The following abbreviations are used in this manuscript:

PIES	Park Integrated Energy Systems
PV	Photovoltaic
LSTM	Long Short-Term Memory
DT	Digital Twin
DRO	Distributionally Robust Optimization
LHS	Latin Hypercube Sampling
RMSE	Root Mean Square Error
MAE	Mean Absolute Error
CCG	Column-And-Constraint Generation
UB	Upper Bound
LB	Lower Bound

References

1. Yue, Y.; Miao, A.; Wu, H.; Zhu, J.; Wang, Z.; Qian, K. Review of the Research Framework and Key Issues for Low-carbon Integrated Energy System. *High Volt. Eng.* **2024**, *50*, 4019–4036. (In Chinese)
2. Zhang, S.; Wang, D.; Cheng, H.; Song, Y.; Yuan, K.; Du, W. Key Technologies and Challenges of Low-carbon Integrated Energy System Planning for Carbon Emission Peak and Carbon Neutrality. *Autom. Electr. Power Syst.* **2022**, *46*, 189–207. (In Chinese)
3. Dong, C.; Wang, Z.; Bai, J.; Jiang, J.; Wang, B.; Liu, G. Review of Ultra-short-term Forecasting Methods for Photovoltaic Power Generation. *High Volt. Eng.* **2023**, *49*, 2938–2951. (In Chinese)
4. Qin, W.; Tang, H.; Ren, M.; Liang, X.; Wang, T.; Chen, T. Improved CGAN Photovoltaic Short-Term Output Scenario Generation Method Based on CNN-LSTM. *J. Sol. Energy* **2025**, *46*, 263–272. (In Chinese)
5. Sun, F.; Li, L.; Bian, D.; Bian, W.; Wang, Q.; Wang, S. Photovoltaic power prediction based on multi-scale photovoltaic power fluctuation characteristics and multi-channel LSTM prediction models. *Renew. Energy* **2025**, *246*, 122866. [CrossRef]
6. Irfan, M.; Shaf, A.; Ali, T.; Zafar, M.; AlThobiani, F.; Almas, M.A.; Attar, H.M.; Alqhatani, A.; Rahman, S.; Almawgani, A.H.M. Global horizontal irradiance prediction for renewable energy system in Najran and Riyadh. *AIP Adv.* **2024**, *14*, 035137. [CrossRef]
7. Wadood, A.; Albalawi, H.; Alatwi, A.M.; Anwar, H.; Ali, T. Design of a Novel Fractional Whale Optimization-Enhanced Support Vector Regression (FWOA-SVR) Model for Accurate Solar Energy Forecasting. *Fractal Fract.* **2025**, *9*, 35. [CrossRef]
8. Shi, K.; Zhang, D.; Han, X.; Xie, Z. Digital Twin Model of Photovoltaic Power Generation Prediction Based on LSTM and Transfer Learning. *Power Syst. Technol.* **2022**, *46*, 1363–1372. (In Chinese)
9. Lai, B.; Hao, J.; Yang, T.; Du, X.; Wang, S.; Lyu, H.; Chen, J. Development and Challenges of Intelligent Operation and Maintenance of Integrated Energy Systems Based on Digital Twin. *Proc. CSEE* **2025**, 1–17. (In Chinese) [CrossRef]
10. Shahmoradi, S.; Imani, M.H.; Mazza, A.; Pons, E. Applications of the Digital Twin and the Related Technologies Within the Power Generation Sector: A Systematic Literature Review. *Energies* **2025**, *18*, 5627. [CrossRef]
11. Hindi, A.T.; Irfan, M.; Yasin, S.; Draz, U.; Ali, T.; Yasin, I.; Rahman, S. Hybrid quantum-inspired proximal policy optimization for fault detection in wind turbine on supervisory control and data acquisition system. *Proc. Inst. Mech. Eng. Part I J. Syst. Control Eng.* **2025**. [CrossRef]
12. Irfan, M.; Yasin, S.; Draz, U.; Ali, T.; Yasin, I.; Kareri, T.; Rahman, S. Revolutionizing wind turbine fault diagnosis on supervisory control and data acquisition system with transparent artificial intelligence. *Int. J. Green Energy* **2025**, *22*, 2029–2045. [CrossRef]
13. Alatwi, A.M.; Albalawi, H.; Wadood, A.; Anwar, H.; El-Hageen, H.M. Deep Learning-Based Dust Detection on Solar Panels: A Low-Cost Sustainable Solution for Increased Solar Power Generation. *Sustainability* **2024**, *16*, 8664. [CrossRef]

14. Wu, X.; Chen, Z.; Jiang, H.; Luo, S.; Zhao, Y.; Zhao, D.; Dang, P.; Gao, J.; Lin, L.; Wang, H. Forecasting to Foresight: Building an Autonomous O&M Brain for the New Power System Based on a Cognitive Digital Twin. *Electronics* **2025**, *22*, 4537. [CrossRef]
15. Jiming, C.; Qian, X.; Yong, L. Optimal Dispatch of Electricity-Natural Gas Interconnection System Considering Source-Load Uncertainty and Virtual Power Plant with Carbon Capture. *J. Sol. Energy* **2023**, *44*, 9–18. (In Chinese)
16. Cui, Y.; Wang, C.; Niu, Y.; Wang, K.; Yao, W.; Liu, C. Two-stage Stochastic Optimization Decision of Integrated Energy System Considering the Uncertainty of Integrated Demand Response. *Power Syst. Technol.* **2025**, *45*, 2232–2242. (In Chinese)
17. Han, X.; Wang, Z.; Li, H.; Sun, S. Stackelberg Game-Robust Optimal Scheduling of Integrated Energy System in Parks Accounting for Multi-Scenario Collaborative Carbon Reduction. *J. Sol. Energy* **2025**, 1–15. (In Chinese) [CrossRef]
18. Gang, D.; Dongmei, Z.; Xin, L. Research Review on Optimal Scheduling Considering Wind Power Uncertainty. *Proc. CSEE* **2023**, *43*, 2608–2626. (In Chinese)
19. Cheng, S.; Lu, Y.; Wang, C. Distributed robust optimal scheduling for a regional integrated energy system considering coordinated operation of P2G-CCS-HFC. *Power Syst. Prot. Control* **2025**, *53*, 87–101. (In Chinese)
20. Shui, Y.; Liu, J.; Gao, H.; Huang, S.; Jiang, Z. A Distributionally Robust Coordinated Dispatch Model for Integrated Electricity and Heating Systems Considering Uncertainty of Wind Power. *Proc. CSEE* **2018**, *38*, 7235–7248. (In Chinese)
21. Grieves, M. *Product Lifecycle Management: Driving the Next Generation of Lean Thinking by Michael Grieves*; Wiley Online Library: Hoboken, NJ, USA, 2007.
22. He, X.; Ai, Q.; Zhu, T.; Qiu, C.; Zhang, D. Opportunities and Challenges of the Digital Twin in Power System Application. *Power Syst. Technol.* **2020**, *44*, 2009–2019. (In Chinese)
23. Chen, C.; Xiong, G.; Fang, H.; Luo, Y. Short-term photovoltaic power interval prediction model based on MODWT-CEEMDAN-LSTM. *J. Sol. Energy* **2025**, *46*, 416–424. (In Chinese)
24. Wang, L.; Li, C.; Liu, J.; Li, C. Short term photovoltaic power prediction based on GRO-SSA-LSTM. *J. Sol. Energy* **2025**, *46*, 401–409. (In Chinese)
25. Nan, B.; Jiang, C.; Dong, S.; Xu, C. Day-ahead and Intra-day Coordinated Optimal Scheduling of Integrated Energy System Considering Uncertainties in Source and Load. *Power Syst. Technol.* **2023**, *47*, 3669–36683. (In Chinese)
26. Zheng, S.; Xu, H.; Lang, J.; Xia, H. Multi-scenario distributed robust optimal scheduling of multi-area integrated energy systems considering photovoltaic uncertainty. *J. Sol. Energy* **2024**, *45*, 460–469. (In Chinese)
27. Wen, X.; He, M.; Zhang, J.; Zhou, K.; Cai, Y.; Zhang, F. Research on ultra-short-term power forecasting of photovoltaic clusters based on K-means clustering. *Power Syst. Prot. Control* **2025**, *53*, 165–172. (In Chinese)
28. Ju, Y.; Kang, X.; Liu, W.; Zhang, J. Distributed Coordinated Day-ahead Scheduling Method for Distribution Network and Microgrid Based on Distributionally Robust Optimization. *Power Syst. Autom.* **2024**, *48*, 48–58. (In Chinese)

energies

Article

Adaptive Rolling-Horizon Optimization for Low-Carbon Operation of Coupled Transportation–Power Systems

Zhe Zhang [1], Shiyan Luan [1], Yingli Wei [1], Fan Tang [1], Haosen Li [1], Pengkun Sun [2] and Chao Yang [2,*]

1 Xiong'an New Area Power Supply Company, State Grid Hebei Electric Power Co., Ltd., Xiong'an 071600, China; xa_zhangz4@he.sgcc.com.cn (Z.Z.); xa_luansy@he.sgcc.com.cn (S.L.); xa_weiyl@he.sgcc.com.cn (Y.W.); xa_tangf@he.sgcc.com.cn (F.T.); xa_lihs@he.sgcc.com.cn (H.L.)

2 Department of Electrical Engineering, North China Electric Power University (Baoding Campus), Baoding 071003, China; 220242213185@ncepu.edu.cn

* Correspondence: yangchao@ncepu.edu.cn

Abstract

The rapid growth of electric vehicles (EVs) has created new challenges for the coordinated low-carbon operation of transportation and power systems. To address this issue, this paper proposes an adaptive rolling-horizon dynamic user equilibrium (DUE) optimization framework for the low-carbon operation of coupled transportation–power systems. The framework integrates transportation, power, and environmental dimensions into a unified objective. On the transportation side, a DUE-based traffic assignment formulation captures both road travel times and station-level queuing dynamics, providing a realistic representation of EV user behavior. This DUE-based traffic assignment model is coupled with an optimal AC power flow formulation to ensure grid feasibility and quantify network losses. To internalize environmental costs, a carbon emission flow module propagates generator-specific carbon intensities to charging stations, aligning charging decisions with their true emission sources. These components are coordinated within a rolling-horizon method in which the prediction window adapts its length to the variability of demand and renewable forecasts. The proposed method allows longer horizons to improve foresight in stable conditions and shorter ones to maintain robustness under volatility. Numerical case studies demonstrate the effectiveness and robustness of the proposed framework and its potential to support low-carbon, high-efficiency operation of coupled transportation–power systems.

Keywords: transportation–power coupled systems; low-carbon operation; dynamic user equilibrium; rolling-horizon optimization

1. Introduction

Urban infrastructures are undergoing a profound transformation toward sustainability, where transportation and power systems are increasingly interwoven. Electric vehicles (EVs) embody this convergence by acting simultaneously as transportation assets and mobile electricity loads. Their rapid deployment not only reshapes travel demand but also introduces dynamic and spatially concentrated loads into power systems. Such dual impacts intensify the interdependence between the two domains, offering opportunities for efficiency gains and emission reductions while at the same time creating new layers of operational complexity. These intertwined challenges underscore the need for integrated approaches that can capture cross-domain interactions and support the sustainable operation of coupled transportation–power systems (CTPSs).

1.1. Motivation

In CTPSs, EV charging demand exhibits strong spatiotemporal variability. On the transportation side, this variability not only alters travel times but also significantly influences waiting times and service reliability at charging stations. On the power side, aggregated charging loads reshape power flow distributions, disrupt nodal balance, and increase system losses and carbon emissions, thereby influencing both system economics and environmental performance. While such bidirectional interactions create opportunities for cross-domain coordination, they also make CTPSs highly sensitive to demand fluctuations and forecast uncertainty [1,2].

Most existing approaches rely on fixed prediction horizons [3]. When the horizon is extended, forecast errors inevitably accumulate over time, whereas when the horizon is shortened, the optimization lacks sufficient foresight [4]. This mismatch restricts the ability to balance long-term benefits with short-term robustness, often leading to suboptimal routing decisions and inefficient charging behaviors that raise overall operating costs. To address these challenges, this paper proposes an adaptive rolling-horizon optimization framework with a DUE-based traffic assignment module, AC-OPF constraints, and carbon emission flow coupling. The proposed framework integrates traffic equilibrium, power system constraints, and carbon emission flows into a unified paradigm. By dynamically adjusting the prediction window according to forecast variability, it achieves a balance between foresight and robustness, thereby providing a practical pathway toward resilient, low-carbon, and efficient operation of CTPSs.

1.2. Related Literature

Research on CTPSs was initially driven by charging infrastructure planning and coordinated operation. Early studies focused on site selection, station sizing, and service levels under traffic demand and distribution system constraints, aiming to mitigate congestion and reduce grid losses [5]. Over time, this stream of work evolved into operational co-optimization, where station choices on the transportation side and power flow feasibility on the grid side were solved jointly. To manage computational complexity, many contributions relied on simplified grid models, such as linear relaxations, second-order cone relaxations, or DC power flow. With the advancement of data availability and computing resources, AC optimal power flow (AC-OPF) has been increasingly adopted to represent voltages, reactive power, and real-power losses more accurately, thereby enabling feasible and economically meaningful operations [6]. However, many of these studies optimized only a single domain, either minimizing travel time [7,8] or grid operating costs [9,10], while environmental costs were often appended via average emission factors [11]. This separation prevented travel time, grid losses, and carbon emissions from being valued on a comparable basis. As a result, cross-domain externalities, such as traffic-induced charging spikes that stress the grid, were frequently underestimated.

From a behavioral perspective, Wardrop's user equilibrium (UE) provides the foundation for static traffic assignment, where travelers select routes such that all used paths between an origin and destination (OD) pair have equal travel times [12–14]. Extending this concept to the temporal domain leads to dynamic user equilibrium (DUE), which enforces consistency over time by capturing flow conservation, inflow–outflow relations, and congestion propagation along bottleneck links. In the context of EVs, charging stations are commonly modeled as "charging arcs", so that route choice simultaneously determines

 https://doi.org/10.3390/en19010227

whether to visit a station and whether to charge there [15]. This formulation enables the joint evaluation of travel times and charging-related delays within a unified equilibrium framework. However, many existing studies adopt simplified assumptions for station-level service performance, which may overlook the complex interactions between charging demand and system-level constraints. Consequently, there remains a need for equilibrium-based models that capture EV users' charging decisions in a manner consistent with both transportation dynamics and power system operations [16].

On the power system side, modeling of flows and losses forms another essential thread of CTPS research [17]. While DC power flow has often been employed for its scalability, AC-OPF provides a more faithful representation by explicitly capturing voltages, reactive power, and system losses, which are critical under high EV penetration. In parallel, environmental considerations have progressed from coarse average emission factors to the carbon emission flow (CEF) framework. CEF traces generator-specific carbon intensities through network topology and allocates them to buses according to power flow distributions, allowing the carbon cost of EV charging at each station to be aligned with its true upstream generation sources. This coupling ensures that charging decisions reflect both operational and environmental costs, avoiding the spatial and temporal biases inherent in average emission factors [18]. Despite these advances, systematic approaches that integrate CEF with DUE-based traffic assignment and AC-OPF constraints remain limited, and few studies have embedded such a coupling into a rolling multi-period framework, which highlights the need for unified cross-domain models.

Uncertainty introduces another critical challenge for CTPS operations, as both travel demand and renewable generation exhibit strong temporal variability. Rolling optimization, or model predictive control (MPC), provides a general paradigm to balance foresight and robustness by solving finite-horizon subproblems with forecasts and updating them as new information arrives [19]. While this idea has been successfully applied to congestion management in transportation systems [20] and renewable integration in power systems [21], its application to CTPSs remains limited. Existing studies [22] typically employ fixed prediction windows, which can reduce multi-period costs under stable conditions but struggle when demand or renewable outputs become volatile. Short horizons lack foresight, whereas long horizons amplify forecast errors across domains [23,24]. These limitations motivate adaptive strategies that adjust the window length according to forecast variability, using longer horizons in stable periods to extend foresight and shorter ones in volatile periods to limit error propagation [25]. However, systematic approaches that incorporate such adaptive horizons into CTPS optimization remain scarce. In particular, there is a lack of methods that can jointly capture travel behavior, power system feasibility, and carbon impacts within a unified rolling framework, which leaves a key gap that this paper seeks to address [26,27].

1.3. Main Contributions

To address the above gaps, this paper develops an adaptive rolling-horizon optimization framework for CTPSs, in which the transportation subsystem is modeled using a DUE-based assignment formulation. The framework simultaneously unifies behavioral, physical, and environmental dimensions and incorporates adaptive rolling-horizon decision-making. The main contributions of this work are summarized as follows:

- ***Unified cross-domain objective for the CTPS:*** A single monetary objective that consistently values transportation and power-side impacts is established, including travel and queuing delays, distribution system losses, and charging-related carbon emissions.

https://doi.org/10.3390/en19010227

This integration enables a balanced assessment of system-wide operating costs and ensures that behavioral, physical, and environmental factors are jointly optimized;

- ***Adaptive rolling-horizon optimization:*** A receding-horizon scheme is developed, in which the prediction window dynamically adjusts to the variability of travel demand and renewable power forecasts. The method extends foresight under stable conditions and mitigates error propagation under volatile conditions, thereby achieving robust multi-period performance, particularly in congestion-intensive scenarios.

The remainder of this paper is organized as follows. Section 2 presents the mathematical formulation of the proposed adaptive rolling-horizon framework. Section 3 presents numerical results and demonstrates the effectiveness of the proposed framework. Section 4 concludes this work.

2. Materials and Methods

This section develops a rolling-horizon optimization framework for coupled transportation–power systems. The system is represented as an integrated model of road traffic flows, power system operations, and charging stations located at traffic nodes, all formulated in discrete time. The framework adopts a receding-horizon scheme to capture multi-period interactions under uncertainty, as shown in Figure 1.

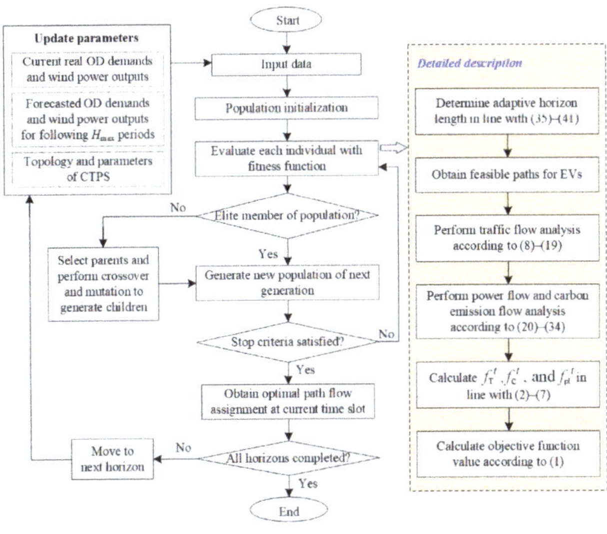

Figure 1. Flowchart of the solution method.

2.1. Objective Function and Cost Decomposition

The overall objective of the rolling-horizon optimization is to minimize the total system cost across the coupled transportation–power system over the prediction horizon. At time step t, the objective function is expressed as

$$\min F_t = \sum_{h=0}^{H_t} \rho^h \left(f_{\mathrm{T}}^{t+h} + f_{\mathrm{pl}}^{t+h} + f_{\mathrm{c}}^{t+h} \right) \quad (1)$$

where $\rho \in (0,1)$ is a discount factor; H_t denotes the prediction window length; and f_{T}^{t+h}, f_{pl}^{t+h}, and f_{c}^{t+h} denote the travel time cost in the transportation system, the power-loss

https://doi.org/10.3390/en19010227

cost in the power system, and carbon emission cost of the EV charging stations at time $t + h$, respectively.

The transportation cost is defined as the sum of on-road driving time cost f^t_{drive} and the queuing delay cost f^t_{station} at charging stations, shown as follows:

$$f^t_{\text{T}} = f^t_{\text{drive}} + f^t_{\text{station}} \tag{2}$$

The driving time component is given by

$$f^t_{\text{drive}} = \sum_{a \in \mathcal{A}_{\text{N}}} \int_0^{d^t_a} c_{\text{time}} t_a(\omega) d\omega \tag{3}$$

where $\mathcal{A}_{\text{N}}$ denotes the set of road arcs, d^t_a denotes the traffic flow on arc a at time t, c_{time} is the monetary value of time, and $t_a(\omega)$ is the travel time on arc a as a function of flow level ω, which captures the flow-dependent congestion effect. The station waiting time cost is modeled as

$$f^t_{\text{station}} = \sum_{a \in \mathcal{A}_{\text{C}}} \int_0^{v_a d^t_a} c_{\text{time}} t^{\text{chg}}_a(\omega) d\omega \tag{4}$$

where $\mathcal{A}_{\text{C}}$ is the set of charging arcs, v_a is the proportion of vehicles choosing to charge on arc a, and $t^{\text{chg}}_a(\omega)$ is the expected waiting time under charging demand ω.

The power-side cost is represented by active power losses in the power system, valued through a monetary coefficient:

$$f^t_{\text{pl}} = c_{\text{pl}} \sum_{(i,j) \in \mathcal{D}_{\text{A}}} G_{ij} \left[(V^t_i)^2 + \left(V^t_j\right)^2 - 2V^t_i V^t_j \cos\theta^t_{ij} \right] \tag{5}$$

where $\mathcal{D}_{\text{A}}$ is the set of branches in power system; G_{ij} is the conductance of branch (i,j); V^t_i and V^t_j are voltage magnitudes of bus i and j, respectively; θ^t_{ij} is the voltage angle difference between buses i and j; and c_{pl} is the unit cost of power loss.

The carbon cost is derived from generator-specific emission intensities, propagated through the CEF module:

$$f^t_{\text{c}} = \sum_{a \in \mathcal{A}_{\text{C}}} \zeta_{a,t} e_{a,t} c_{\text{carb}} \tag{6}$$

where $\zeta_{a,t}$ is the carbon intensity allocated to charging arc a at time t, c_{carb} is the unit cost of carbon emissions, and $e_{a,t}$ is the energy consumption of charging arc a at time t, which can be calculated as follows:

$$e_{a,t} = P^{\text{ch}}_{a,t} \Delta t \tag{7}$$

where $P^{\text{ch}}_{a,t}$ denotes the charging power on arc a at time t, and Δt represents the length of the time interval.

In summary, the above formulation decomposes the objective into transportation, power, and carbon emission components, which are valued on a unified monetary basis. This structure ensures that routing decisions, power system operations, and environmental impacts are jointly considered within the rolling-horizon optimization framework.

2.2. *Traffic System Constraints*

The transportation system determines both travel delays on roads and queuing delays at charging stations, which together shape the spatiotemporal distribution of EV charging demand. To capture these dynamics in a consistent and tractable manner, we adopt a graph-based representation of the road network combined with an M/M/c queuing model

https://doi.org/10.3390/en19010227

(Erlang-C formula) for charging stations. This approach ensures that route choices, station arrivals, and service delays are coherently linked within the optimization framework.

The travel time on each road arc $a \in \mathcal{A}_{\mathrm{N}}$ is modeled by the classical Bureau of Public Roads (BPR) function [5,28]:

$$t_a(d_a^t) = t_a^0\left[1 + 0.15\left(\frac{d_a^t}{c_a}\right)^4\right], \ \forall a \in \mathcal{A}_{\mathrm{N}} \tag{8}$$

where t_a^0 is the free-flow travel time of road arc a, and c_a denotes road capacity of arc a. This nonlinear function captures the congestion effect as flows approach capacity. The transportation network topology considered in this work is illustrated in Figure 2b, and the corresponding link-specific parameters used in the BPR function, including the free-flow travel time t_a^0 and road capacity c_a, are summarized in Table 1.

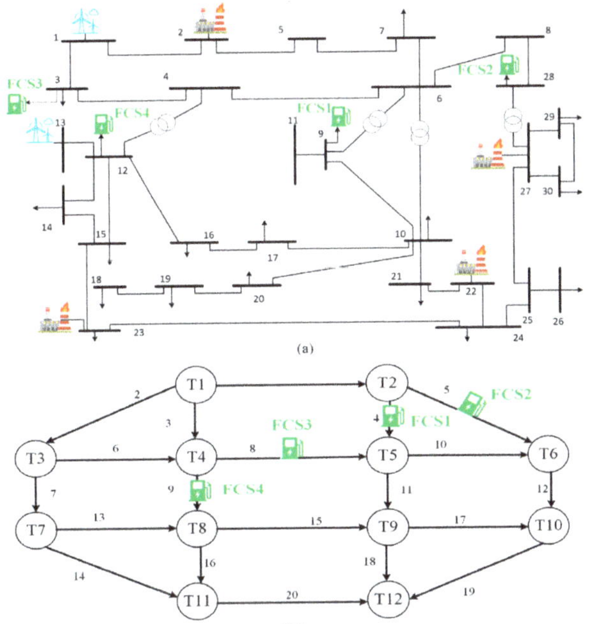

Figure 2. Network topology of the test coupled transportation–power systems. (**a**) Power system network. (**b**) Transportation system network.

Table 1. Transportation system parameters.

Road	c_a (p.u.)	t_a^0 (h)	Road	c_a (p.u.)	t_a^0 (h)	Road	c_a (p.u.)	t_a^0 (h)
T1–T2	16.8	0.60	T4–T5	12.6	0.75	T7–T11	14.7	0.30
T1–T3	14.0	0.30	T4–T8	14.0	0.66	T8–T11	11.2	0.24
T1–T3	12.6	0.30	T5–T6	13.3	0.45	T9–T10	13.3	0.30
T2–T5	12.6	0.30	T5–T9	14.0	0.75	T9–T12	11.2	0.24
T2–T6	13.3	0.54	T7–T8	14.0	0.30	T6–T10	13.3	0.60
T3–T4	11.2	0.30	T8–T9	14.0	0.66	T10–T12	10.5	0.30
T3–T7	14.0	0.30				T11–T12	15.4	0.54

https://doi.org/10.3390/en19010227

For each charging arc $a \in \mathcal{A}_C$, the arrival rate is obtained as

$$\lambda_a^t = \frac{x_a^t}{\Delta t}, \forall a \in \mathcal{A}_C \tag{9}$$

where x_a^t is the number of vehicles arriving at charging arc a in time slot t. The charging station utilization factor is defined as

$$\psi_a^t = \frac{\lambda_a^t}{c_a^{\mathrm{CF}} \mu_a}, \forall a \in \mathcal{A}_C \tag{10}$$

where c_a^{CF} denotes the number of chargers at charging station a, and μ_a is the service rate of each charger. To ensure stability, the following condition must be held, which means that the arrival rate cannot exceed the aggregate service capacity.

$$0 \leq \psi_a^t < 1, \forall a \in \mathcal{A}_C \tag{11}$$

Queueing at a charging station is captured by the multi-server Erlang-C formula. The empty-system probability of the M/M/c queue is given by

$$P_{0,a}^t = \left[\sum_{n=0}^{c_a^{\mathrm{CF}}-1} \frac{\left(c_a^{\mathrm{CF}} \psi_a^t\right)^n}{n!} + \frac{\left(c_a^{\mathrm{CF}} \psi_a^t\right)^{c_a^{\mathrm{CF}}}}{c_a^{\mathrm{CF}}!(1-\psi_a^t)} \right]^{-1}, a \in \mathcal{A}_C \tag{12}$$

where $P_{0,a}^t$ denotes the probability that charging station a is empty at the beginning of slot t. This term also serves as the normalization constant for the steady-state distribution of the M/M/c queue. Based on the Erlang-C model, the expected waiting time per arriving EV at charging station a is given by

$$W_a^t = \frac{P_{0,a}^t \left(\frac{\lambda_a^t}{\mu_a}\right)^{c_a^{\mathrm{CF}}} \psi_a^t}{c_a^{\mathrm{CF}}!(1-\psi_a^t)^2 \lambda_a^t}, \quad \forall a \in \mathcal{A}_C \tag{13}$$

Accordingly, the expected service delay on charging arc a is given by

$$t_a^{\mathrm{chg}}(d_a^t) = \frac{1}{\mu_a} + W_a^t, \forall a \in \mathcal{A}_C \tag{14}$$

where $1/\mu_a$ is the average charging time per vehicle, and the second term accounts for additional waiting due to congestion at the station.

Because vehicles continue along their chosen paths once they reach a node, the available departures for OD (r, s) in slot t equal the exogenous demand plus the outflows from all incoming links to the origin node:

$$\sum_{k \in K_{rs}} f_k^{rs,t} = q_{rs}^t + \sum_{a \in \mathcal{A}_{\mathrm{ex}}(r)} O_a^{s,t}, \forall r, s, t \tag{15}$$

where K_{rs} denotes the set of all feasible paths between origin r and destination s, and let $k \in K_{rs}$ index a specific feasible path. Let $f_k^{rs,t}$ denote the flow on path k for OD pair (r, s) at time t, and q_{rs}^t the travel demand of OD pair (r, s) at time t. $\mathcal{A}_{\mathrm{ex}}(r)$ is the set of arcs with an

https://doi.org/10.3390/en19010227

exit node r, $O_a^{s,t}$ is the outflow on link a in slot t bound for destination s, and total outflow constraints are imposed in (16).

$$O_a^t = \sum_{s \in S} O_a^{s,t}, \forall a, \forall t \tag{16}$$

This aggregates destination-wise outflows into the total outflow from link a. Path–link incidence is given as follows:

$$x_a^t = \sum_{r \in R} \sum_{s \in S} \sum_{k \in K_{rs}} f_k^{rs,t} \delta_{k,a}^{rs}, \forall a, \forall t \tag{17}$$

where R and S denote the sets of origins and destinations, respectively, and $\delta_{k,a}^{rs} \in \{0,1\}$ indicates whether link a lies on path k, which maps path flows to link arrivals x_a^t. It denotes the flow newly assigned to this arc at time t from the OD demands under the user-equilibrium model; accordingly, the traffic flow on this arc at time t is expressed as

$$d_a^t = d_a^{t-1} + x_a^t - O_a^t, \forall a, \forall t \geq 1 \tag{18}$$

$$d_a^0 = 0, \forall a, \forall t \tag{19}$$

It is worth noting that the DUE formulation adopted in this study is based on the standard assumption of fully rational and fully informed EV users, who select routes and charging stations to minimize their perceived generalized travel cost. This assumption is widely used in equilibrium-based traffic modeling and enables a tractable representation of collective user behavior. However, it may not fully capture bounded rationality, preference heterogeneity, or information imperfections that can arise in real-world EV routing and charging decisions.

2.3. Power System Constraints and Carbon-Intensity Mapping

We model the power system as a directed graph $(\mathcal{D}_\mathrm{N}, \mathcal{D}_\mathrm{A})$, where $\mathcal{D}_\mathrm{N}$ is the set of buses, and $\mathcal{D}_\mathrm{A}$ is the set of branches. In time slot t, the active and reactive net injections at bus j are given as

$$P_j^t = P_{\mathrm{g},j}^t - P_{\mathrm{l},j}^t - P_{\mathrm{ch},j}^t, \forall j \in \mathcal{D}_\mathrm{N} \tag{20}$$

$$Q_j^t = Q_{\mathrm{g},j}^t - Q_{\mathrm{l},j}^t, \forall j \in \mathcal{D}_\mathrm{N} \tag{21}$$

where $P_{\mathrm{g},j}^t$ and $Q_{\mathrm{g},j}^t$ are the active and reactive outputs of generators, respectively; $P_{\mathrm{l},j}^t$ and $Q_{\mathrm{l},j}^t$ are the active and reactive power demands, respectively; and $P_{\mathrm{ch},j}^t$ is the aggregate EV charging power at bus j. The latter is obtained from the charging arcs connected to bus j:

$$P_{\mathrm{ch},j}^t = \sum_{a \in \mathcal{A}_\mathrm{C}(j)} \left(v_a d_a^t P^{\mathrm{EV}} \right), \forall j \in \mathcal{D}_\mathrm{N} \tag{22}$$

where $\mathcal{A}_\mathrm{C}(j) \subseteq \mathcal{A}_\mathrm{C}$ is the set of charging stations electrically connected to bus j, and each station corresponds to a charging arc on the traffic side; P^{EV} is the per vehicle charging power.

To represent active and reactive power conservation in the power system, we employ branch-level balance equations. For each branch $(i, j) \in \mathcal{D}_\mathrm{A}$, the sending-end power equals the sum of the nodal injection at bus j, the power flows dispatched to its downstream branches, and the power losses on branch (i, j). This formulation captures both power

https://doi.org/10.3390/en19010227

losses and flow redistribution, ensuring that the system representation remains consistent with AC power flow physics:

$$P_{ij}^t = P_j^t + R_{ij} l_{ij}^t + \sum_{k:(j,k) \in \mathcal{D}_\mathrm{A}} P_{jk}^t, \forall (i,j) \in \mathcal{D}_\mathrm{A} \tag{23}$$

$$Q_{ij}^t = Q_j^t + X_{ij} l_{ij}^t + \sum_{k:(j,k) \in \mathcal{D}_\mathrm{A}} Q_{jk}^t, \forall (i,j) \in \mathcal{D}_\mathrm{A} \tag{24}$$

with voltage drop and current–power coupling

$$\left(V_j^t\right)^2 = (V_i^t)^2 - 2(R_{ij} P_{ij}^t + X_{ij} Q_{ij}^t) + (R_{ij}^2 + X_{ij}^2) l_{ij}^t \tag{25}$$

where P_{ij}^t and Q_{ij}^t are the active and reactive branch flows, respectively; R_{ij} and X_{ij} are branch resistance and reactance, respectively; and l_{ij}^t is the squared current magnitude on branch (i,j).

In addition, the thermal rating of each branch and voltage magnitude at each node must be maintained within the allowable range, given as follows:

$$\sqrt{(P_{ij}^t)^2 + (Q_{ij}^t)^2} \leq S_{ij}^{\max}, \forall (i,j) \in \mathcal{D}_\mathrm{A} \tag{26}$$

$$V_i^{\min} \leq V_i^t \leq V_i^{\max}, \quad \forall i \in \mathcal{D}_\mathrm{N} \tag{27}$$

where $\left[V_i^{\min}, V_i^{\max}\right]$ are voltage magnitude limits, and $S_{ij}^{\max}$ is the thermal rating of branch (i,j). The active and reactive power outputs of each generator must satisfy their respective operating limits. These constraints are formulated as follows:

$$P_{\mathrm{g},i}^{\min} \leq P_{\mathrm{g},i}^t \leq P_{\mathrm{g},i}^{\max}, \quad \forall i \in \mathcal{D}_\mathrm{N} \tag{28}$$

$$Q_{\mathrm{g},i}^{\min} \leq Q_{\mathrm{g},i}^t \leq Q_{\mathrm{g},i}^{\max}, \quad \forall i \in \mathcal{D}_\mathrm{N} \tag{29}$$

where $\left[P_{\mathrm{g},i}^{\min}, P_{\mathrm{g},i}^{\max}\right]$ and $\left[\ Q_{\mathrm{g},i}^{\min}, Q_{\mathrm{g},i}^{\max}\right]$ represent the active and reactive generation bounds, respectively.

To discourage excessive voltage excursions, we introduce an auxiliary variable χ_i^t that measures the per unit voltage deviation at bus i. The deviation is constrained within an admissible tolerance:

$$\chi_i^t = \frac{V_i^t - V_{i,\mathrm{ref}}}{V_{i,\mathrm{ref}}}, \quad \forall i \in \mathcal{D}_\mathrm{N} \tag{30}$$

$$\left|\chi_i^t\right| \leq \chi_{\max}, \quad \forall i \in \mathcal{D}_\mathrm{N} \tag{31}$$

where $V_{i,\mathrm{ref}}$ denotes the nominal reference voltage at bus i, typically set to the rated value (e.g., 1.0 p.u.), and $\chi_{\max}$ is the admissible tolerance for per unit voltage deviations, ensuring that bus voltages remain within an acceptable fluctuation band.

To prevent transformer overloading and ensure that the apparent power at each bus remains within the rated capacity, we define a loading factor α_i^t. This constraint not only guarantees operational security but also promotes a more balanced utilization of charging stations across the system.

$$\alpha_i^t = \frac{\sqrt{\left(P_{\mathrm{l},i}^t + P_{\mathrm{ch},i}^t\right)^2 + \left(Q_{\mathrm{l},i}^t\right)^2}}{S_i^{\mathrm{TR}}}, \forall i \in \mathcal{D}_\mathrm{N} \tag{32}$$

https://doi.org/10.3390/en19010227

$$\alpha_i^t \leq \alpha_{\max}, \forall i \in \mathcal{D}_{\mathrm{N}} \tag{33}$$

where S_i^{TR} denotes the rated transformer capacity at node i, α_i^t represents the loading factor at node i in time slot t, and $\alpha_{\max}$ specifies the admissible maximum loading factor.

We adopt a CEF mapping to propagate generator-specific carbon intensities through the system to the consumption points (charging buses). Let $\zeta_{a,t}$ denote the carbon intensity allocated to charging arc a at time t. The carbon intensity that feeds the charging station located at bus i is computed as the ratio of carbon-attributed supply to total supply at that bus:

$$\zeta_{i,t} = \frac{\sum\limits_{h\in\Omega_i^{\mathrm{coal}}} P_{h,t}^{\mathrm{coal}} \cdot \zeta_h^{\mathrm{coal}} + \sum\limits_{ij\in\Omega_i^L} P_{ij}^t \cdot \zeta_{ij,t}^{\mathrm{line}}}{\sum\limits_{h\in\Omega_i^{\mathrm{coal}}} P_{h,t}^{\mathrm{coal}} + \sum\limits_{y\in\Omega_i^{\mathrm{wind}}} P_{y,t}^{\mathrm{wind}} + \sum\limits_{ij\in\Omega_i^L} P_{ij}^t} \tag{34}$$

where Ω_i^{coal}, Ω_i^{wind}, and Ω_i^L denote the sets of coal units, wind power units, and branches that directly inject active power into bus i, respectively; $P_{h,t}^{\mathrm{coal}}$ and $P_{y,t}^{\mathrm{wind}}$ are the outputs of thermal and wind power units, respectively; ζ_h^{coal} and $\zeta_{ij,t}^{\mathrm{line}}$ are the carbon emission intensities of coal unit h and branch ij, respectively; and P_{ij}^t is the active power on branch ij.

2.4. Determination of Adaptive Horizon Length

To allow longer horizons that enhance foresight under stable conditions and shorter ones that preserve robustness under volatile conditions, the rolling-horizon length is adaptively adjusted according to the variability of traffic demand and wind power generation forecasts. Two stability indices are therefore introduced to quantify the temporal fluctuations of OD demand and wind power output, respectively. These indices serve as quantitative signals for shrinking or extending the optimization horizon, ensuring that the framework dynamically adapts to different levels of forecast uncertainty.

For each OD pair rs, the current real demand at time t, denoted by q_{rs}^t, and the forecasted values for the following $H_{\max}$ periods $\left(q_{rs}^{t+1}, \cdots, q_{rs}^{t+H_{\max}}\right)$ are collected. The maximum among these values is given as

$$q_{rs}^{\max} = \max\left\{q_{rs}^t, q_{rs}^{t+1}, \cdots, q_{rs}^{t+H_{\max}}\right\} \tag{35}$$

Each value is normalized as follows:

$$\widetilde{q}_{rs}^{\tau} = \frac{q_{rs}^{\tau}}{q_{rs}^{\max} + \varepsilon}, \qquad \tau \in \{t, \quad t+1, \quad \cdots, \quad t+H_{\max}\} \tag{36}$$

where ε is a sufficiently small positive constant to avoid division by zero.

The normalized sequence $\{\widetilde{q}_{rs}^{\tau}\}$ lies within [0, 1], and its standard deviation is denoted as σ_{rs}^t. Since the maximum possible standard deviation of values in [0, 1] is 0.5, the OD demand stability index is defined as follows:

$$U_{\mathrm{OD}}^t = \frac{1}{N_{\mathrm{OD}}} \sum_{rs} (1 - 2\sigma_{rs}^t) \tag{37}$$

where N_{OD} is the number of OD pairs.

https://doi.org/10.3390/en19010227

For each wind power unit i, the current real output $P_{i,t}^{\text{wind}}$ and the forecasts $\left(P_{i,t+1}^{\text{wind}}, \ \ldots, \ P_{i,t+H_{\max}}^{\text{wind}}\right)$ are normalized by the rated capacity $P_{\text{rate, }i}^{\text{wind}}$.

$$\widetilde{P}_{i,\tau}^{\text{wind}} = \frac{P_{i,\tau}^{\text{wind}}}{P_{\text{rate, }i}^{\text{wind}}}, \qquad \tau \in \{t, \ t+1, \ \ldots, \ t+H_{\max}\} \tag{38}$$

The standard deviation of normalized outputs is denoted as $\sigma_{\text{wind},i}^{t}$, and the wind power stability index is given as

$$U_{\text{wind}}^{t} = \frac{1}{N_{\text{wind}}} \sum_{i} \left(1 - 2\sigma_{\text{wind},i}^{t}\right) \tag{39}$$

where N_{wind} is the number of wind power units.

The OD and wind stability indices are combined as

$$U_t = \beta U_{\text{OD}}^{t} + (1-\beta) U_{\text{wind}}^{t} \tag{40}$$

where the parameter $\beta \in [0,1]$ controls the relative importance between traffic demand variability and renewable generation variability in determining the adaptive horizon length. A larger β places more emphasis on OD demand fluctuations, leading to a horizon that responds more strongly to traffic congestion dynamics, whereas a smaller β increases the influence of wind power variability and makes the horizon more sensitive to renewable uncertainty. In this study, β is set to 0.5 to balance these two sources of uncertainty, and moderate variations of β were observed to affect the horizon length smoothly without altering the qualitative behavior of the adaptive strategy. Finally, the adaptive horizon length is computed as

$$H_t = H_{\min} + \lfloor (H_{\max} - H_{\min}) U_t + 0.5 \rfloor \tag{41}$$

where $H_{\min}$ and $H_{\max}$ are the lower and upper bounds of the horizon length, and $\lfloor \cdot \rfloor$ denotes the floor operator, which effectively performs rounding when adding 0.5. This ensures that the horizon length shrinks when variability is high (low stability) and extends when conditions are stable, achieving a balance between robustness and foresight.

2.5. Solution Method

The overall solution process of the proposed adaptive rolling-horizon DUE optimization framework is illustrated in Figure 1. At each decision period, the solution method begins with the collection of input data, including current OD demands and wind power outputs together with their multi-step forecasts, as well as the topology and parameters of the CTPS. Based on these inputs, two stability indices are computed to quantify the variability of traffic demand and renewable generation. According to (35)–(41), the adaptive prediction window length is then determined, ensuring that a longer horizon is selected under stable conditions to improve foresight, while a shorter horizon is adopted when variability is high to limit the influence of forecast errors.

Given the prediction window, the feasible path set for EVs is generated, and an initial population of candidate path-flow vectors is established. The optimization follows an elitist genetic algorithm framework, where each individual in the population is evaluated through a fitness function representing the total operating cost of the system. Specifically, for each candidate solution, the inner evaluation pipeline involves solving the dynamic user equilibrium problem with the SQP method to assign traffic flows and

 https://doi.org/10.3390/en19010227

capture in-station queuing dynamics, followed by AC power flow and carbon emission flow analysis to evaluate grid feasibility, system losses, and nodal carbon intensities. The corresponding transportation cost, power loss cost, and carbon cost are calculated according to Equations (2)–(7) and aggregated into the objective function (1) over the prediction horizon using the discount factor.

The genetic algorithm iteratively selects elite members of population, applies crossover and mutation to generate offspring, and updates the population until the stopping criteria are satisfied. Once convergence is reached, the optimal path-flow assignments in current horizon can be obtained. Only the first-period decision is implemented in practice, after which the horizon is rolled forward with updated real-time information and new forecasts. This receding-horizon procedure continues until all horizons have been processed, thereby producing a dynamic sequence of path-flow assignments that jointly balance foresight and robustness across transportation and power domains.

It is worth clarifying that the proposed rolling-horizon optimization framework is not itself a DUE model. Instead, the dynamic user equilibrium is embedded as the traffic assignment component within each rolling-horizon decision step, while the rolling-horizon mechanism serves as an outer-layer framework to coordinate multi-period decisions under forecast uncertainty.

Although the case study considers a medium-scale coupled transportation–power system, the proposed rolling-horizon framework is inherently scalable. The computational burden mainly depends on the number of traffic nodes and candidate paths in the DUE problem, the number of buses and branches in the AC power flow, and the length of the rolling horizon. At each decision step, the dominant computational cost arises from repeatedly solving the embedded traffic assignment and power system evaluation problems. In large-scale applications, the computational efficiency can be further improved through parallel evaluation of candidate solutions, path set reduction, and decomposition techniques, which makes the proposed framework applicable to larger systems.

2.6. Experimental Settings

As shown in Figure 2, the test system integrates a transportation system and a power system to form a tested CTPS. The power system contains 30 buses (T1–T30) and a portfolio of generating units, including four coal-fired plants and two wind power units with distinct output limits and emission factors. It is worth noting that the type of generation unit plays a critical role in the proposed framework, as different generation technologies are associated with distinct carbon emission intensities. These generator-specific emission factors are explicitly considered in the CEF module and directly affect the carbon cost component in the objective function. There are four nodes (T3, T9, T12, and T28) that host fast-charging stations labeled FCS1–FCS4. The transportation system is represented by 12 nodes and 20 directed links. Each charging station is modeled as a charging arc in the transportation system and electrically mapped to its associated bus in the power system, such that charging flows determined by EV route choices translate into active power injections at the corresponding buses. This formulation ensures traffic-induced charging demand is consistently reflected in nodal loads, system losses, and carbon intensities, thereby establishing a realistic physical coupling between two infrastructures.

All simulations were conducted on a standard workstation equipped with an Intel Core i7 processor (2.9 GHz) and 32 GB RAM. For the tested coupled transportation–power system, the average computation time per rolling-horizon step was on the order of tens of seconds. Specifically, the dominant computational cost arises from the repeated solution of the embedded DUE assignment and AC power flow within the GA evaluation

 https://doi.org/10.3390/en19010227

process, while the outer-loop GA convergence typically requires a limited number of generations. These results indicate that the proposed framework is computationally tractable for medium-scale systems and suitable for near-real-time operational studies.

Table 1 presents the fundamental parameters of the transportation system, detailing the 20 directed road segments that connect the 12 nodes. For each segment, the table reports the road capacity and the corresponding free-flow travel time, thereby providing the basis for modeling congestion dynamics under varying traffic conditions. Table 2 specifies the configuration of the power system and generation portfolio, including the bus locations of the generating units, their types, and the associated lower and upper operating limits. In addition, the table differentiates the carbon emission factors across coal-fired and renewable units, which enables a consistent mapping of generator-specific emission intensities to nodal carbon costs in the coupled analysis.

Table 2. Power system and generation portfolio.

Gen ID	Bus	Type	P_{min} (MW)	P_{max} (MW)	ζ (kgCO$_2$/kWh)
1	1	wind	0	80	0
2	2	coal	0	80	0.85
3	13	wind	0	40	0
4	22	coal	0	50	0.85
5	23	coal	0	30	0.85
6	27	coal	0	55	0.85

As presented in Tables 3 and 4, the dataset specifies, for each time period, the realized OD pair demands in the transportation system and the actual wind power outputs in the power system, respectively. The economic parameters are set as follows: the monetary value of travel time is 1 \$/h, the carbon price is 20 \$/tonCO$_2$, and the penalty on active power losses is 25 \$/MW. The rated charging power of each EV is assumed to be 200 kW.

Table 3. Realized wind power output across time periods.

Gen ID	P_{min} (MW)					
	$t = 1$	$t = 2$	$t = 3$	$t = 4$	$t = 5$	$t = 6$
1	10	0	10	20	40	40
3	10	30	0	20	40	40

Table 4. Realized OD demand across time periods.

OD (r–s)	q_{rs} (p.u.)					
	$t = 1$	$t = 2$	$t = 3$	$t = 4$	$t = 5$	$t = 6$
(1–6)	10	120	10	150	120	100
(1–10)	0	0	40	0	0	0
(1–11)	30	60	0	80	50	60
(1–12)	0	100	0	90	60	90
(2–12)	8	20	50	10	10	0
(3–6)	30	0	4	11	0	10
(3–10)	0	0	0	12	12	12
(3–11)	0	100	60	40	40	40
(3–12)	25	0	0	20	20	20
(4–10)	0	30	0	30	30	30
(4–12)	20	0	7	60	40	60

https://doi.org/10.3390/en19010227

3. Results

3.1. Effectiveness of Adaptive Horizon Strategy

To validate the effectiveness of the proposed rolling-horizon optimization, we benchmark three strategies for path-flow assignment: (i) a no-forecast strategy, (ii) a fixed-horizon strategy, and (iii) the proposed adaptive horizon strategy. These three settings differ in how far ahead the optimizer "looks" when computing the operation cost of CTPS. In the no-forecast strategy, the optimizer only considers realized OD demands and wind generation in the current period, without anticipating future conditions. In the fixed-horizon strategy, the optimizer always looks ahead two additional periods, so that decisions for the current period are based on the realized values plus forecasts for the following two periods. To mitigate the effect of forecast error accumulation, later periods are assigned smaller weights in the aggregated objective. Finally, in the adaptive horizon strategy, the prediction horizon length is determined dynamically from the realized and forecasted values: when demand and generation are volatile, the horizon shortens to avoid relying on unreliable long-term forecasts; when conditions are stable, the horizon extends to capture more foresight. This adaptive rule allows the optimization framework to balance robustness and proactivity in real time.

Figure 3 shows the per period total costs obtained under the three horizon strategies. In period 1, all three strategies yield relatively low costs, with only marginal differences across strategies. Starting from period 2, however, the gaps widen significantly. The no-forecast strategy incurs the highest cost of USD 9410, whereas the fixed-horizon and adaptive horizon strategies reduce it to USD 8590 and USD 8580, respectively. In period 3, the difference between the three strategies narrows again, with total costs of USD 5140, USD 5070, and USD 4970. The most pronounced contrast appears in period 4: the no-forecast strategy reaches USD 12,000, the fixed-horizon strategy reduces this to USD 11,600, and the adaptive horizon strategy achieves the lowest value of USD 11,070. These trajectories highlight two important phenomena. First, although forecasting-based strategies may slightly increase cost in the first period compared with the no-forecast baseline, they deliver consistent savings in subsequent periods by anticipating congestion and reallocating flows in advance. Second, the adaptive horizon achieves the lowest cost overall, particularly in the most congested period (period 4), where its dynamic adjustment mechanism provides a clear advantage. Summed across all periods, the cumulative total cost of the no-forecast strategy is about USD 29,160, compared with USD 27,960 under the fixed horizon and USD 27,290 under the adaptive horizon, corresponding to reductions of approximately 4.1% and 6.4%, respectively.

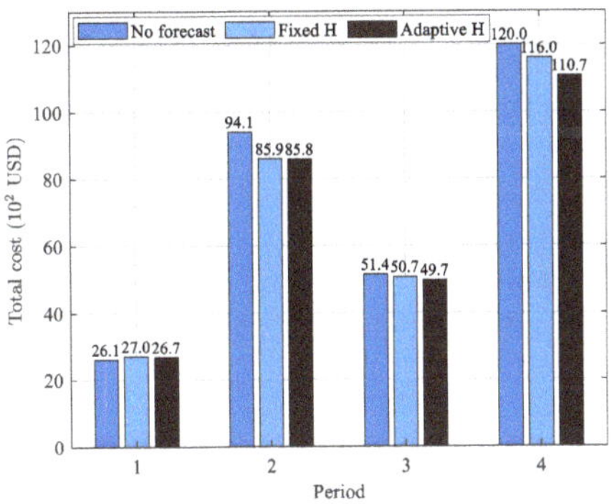

Figure 3. Per period total cost under three horizon strategies.

https://doi.org/10.3390/en19010227

To examine the sources of these savings, Table 5 reports the per period cost breakdown into traffic, grid loss, and carbon emission components. The traffic time cost dominates the total in all strategies, accounting for more than 70% of the overall expenditure. Accordingly, the most substantial reductions are observed in the traffic component, especially in periods 2 and 4 when OD demand is highest. Compared with the no-forecast baseline, the fixed-horizon strategy reduces traffic costs by 12.8%, 1.5%, and 2.1% in periods 2–4, while the adaptive horizon strategy achieves reductions of 9.8%, 4.1%, and 7.3%, respectively. These results confirm that the anticipatory "peak-shaving" effect of forecasting-based strategies effectively suppresses downstream congestion.

Table 5. Per period cost breakdown under three horizon strategies.

Strategy	Item (10^2 USD)	$t = 1$	$t = 2$	$t = 3$	$t = 4$
No forecast	Traffic cost	18.74	78.01	38.74	103.62
	Power loss	5.14	4.64	5.83	5.15
	Carbon cost	2.22	11.44	6.83	8.56
	Total cost	26.11	94.10	51.41	120.04
Fixed H	Traffic cost	19.18	67.98	38.17	101.49
	Power loss	5.18	4.71	5.82	5.22
	Carbon cost	2.71	11.42	6.76	9.34
	Total cost	27.08	85.91	50.76	116.07
Adaptive H	Traffic cost	18.93	70.39	37.16	96.09
	Power loss	5.17	4.69	5.82	5.23
	Carbon cost	2.63	10.79	6.77	8.43
	Total cost	26.74	85.88	49.75	109.76

Beyond the traffic component, the table also reveals two additional insights. First, the power loss costs remain relatively stable across both periods and strategies, fluctuating within a narrow band between USD 4.6×10^2 and USD 5.9×10^2. This stability suggests that transmission losses are less sensitive to horizon design, as they are mainly governed by physical grid constraints. Nonetheless, the no-forecast case shows a slight spike in period 3 (USD 5.83×10^2), while the forecasting-based strategies avoid such increases, indicating that improved flow allocation indirectly stabilizes system operation. Second, carbon emission costs exhibit clearer divergence across strategies in later periods. While differences are negligible in the first period, the adaptive horizon attains the lowest carbon cost in period 4 (USD 8.43×10^2), compared with USD 8.56×10^2 under no-forecast and USD 9.34×10^2 under the fixed horizon. Although the contribution of carbon cost to the total remains modest (below 10%), this consistent downward trend highlights the environmental co-benefits of anticipatory allocation: by guiding EV charging toward less carbon-intensive supply conditions, forecasting strategies reduce not only congestion but also system-wide emissions.

The congestion patterns underlying these results are illustrated in Figure 4, which shows the spatiotemporal evolution of link flow-to-capacity ratios under the three horizon strategies. Green denotes uncongested links (ratio < 0.3), yellow and orange represent moderate congestion (0.3–1.2), and red indicates severe congestion (ratio > 1.2).

https://doi.org/10.3390/en19010227

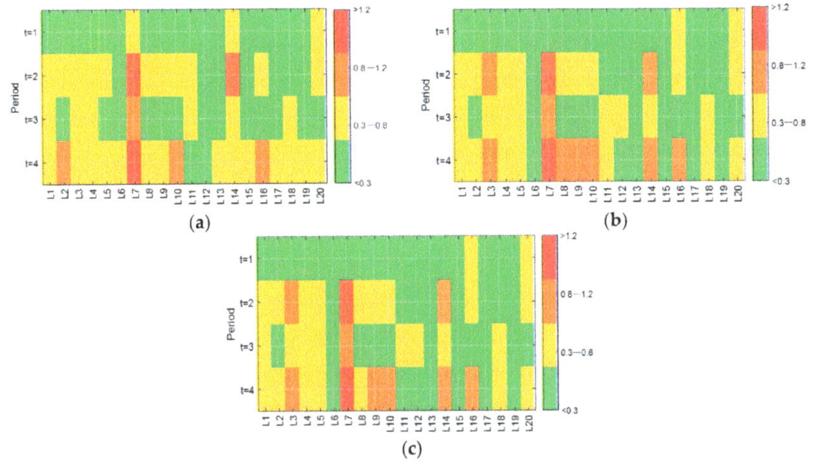

Figure 4. Spatiotemporal road congestion under three horizon strategies. (**a**) No forecast. (**b**) Fixed horizon. (**c**) Adaptive horizon.

Congestion intensity varies considerably across periods. Periods 1 and 3 remain largely uncongested, whereas widespread congestion arises in periods 2 and 4, coinciding with the peak OD demands reported in Table 3. This temporal correspondence highlights how traffic fluctuations drive system stress and underscores the importance of horizon design in managing demand surges to avoid cumulative spillovers. The spatial distribution of congestion reveals that only a few bottleneck links, notably L6, L7, L8, and L13, experience recurrent overload across strategies. Their limited capacity and structural role in connecting major OD pairs make them particularly vulnerable. The forecasting-based strategies relieve these bottlenecks by diverting flows to alternative paths, spreading traffic more evenly across the system. In contrast, the no-forecast baseline allows residual flows to accumulate, producing persistent bottlenecks and a proliferation of red links, especially in periods 2 and 4.

When comparing the two forecasting strategies, the adaptive horizon displays a distinct advantage in the later stages. Although its patterns appear broadly similar to the fixed horizon in periods 1–3, the adaptive horizon results in fewer severely congested links and lower flow magnitudes in period 4. This outcome reflects the benefit of dynamically adjusting horizon length, which suppresses long-run spillovers more effectively than a fixed look ahead.

3.2. Sensitivity Analysis of the Discount Factor

Figure 5 depicts the four-period cumulative total cost as a function of the discount factor ρ under the Fixed-H and Adaptive-H strategies. Both curves display a broadly similar shape: the cost decreases as ρ increases and stabilizes into a plateau from around $\rho \approx 0.5$ onward. At this near-optimal region, the cumulative cost converges to approximately 2730 ($\times 10^2$ USD) for Adaptive H and 2798 ($\times 10^2$ USD) for Fixed H. Further raising ρ toward unity produces negligible additional savings, and the Adaptive H curve remains almost flat with only minor fluctuations. In contrast, when ρ drops below 0.5, the total cost increases rapidly, with especially sharp rises at $\rho = 0.4$ and below. This pattern suggests that small ρ values overweight immediate-period outcomes, undermining anticipatory adjustments and inducing myopic assignments that push congestion and charging costs into later

 https://doi.org/10.3390/en19010227

periods. By comparison, larger ρ values enhance foresight, but beyond about $\rho \approx 0.5$–0.6, the marginal benefits diminish due to forecast-error noise, leading to a robust plateau.

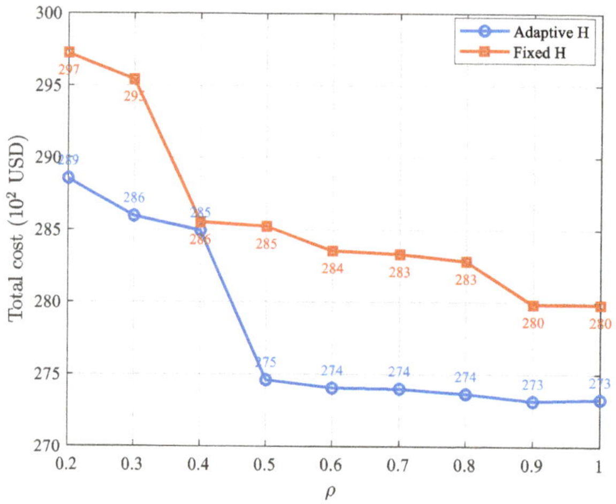

Figure 5. Sensitivity of total cost to the discount factor ρ.

Across the entire range of ρ, the Adaptive H curve consistently lies below the Fixed H curve, confirming the efficiency advantage of dynamically tuning horizon length to volatility. Quantitatively, at $\rho = 0.9$, Adaptive H reduces the cumulative cost by about 2.3% relative to Fixed H; at $\rho = 0.6$, the reduction is approximately 3.3%. Even under the unfavorable low-ρ case ($\rho = 0.2$), Adaptive H still maintains an advantage of around 2.9%, demonstrating robustness against the choice of discount factor. Mechanistically, the effect of ρ operates through two main channels: (i) adjusting the intertemporal weights in the objective, thereby promoting earlier peak shaving and upstream flow diversion to reduce downstream congestion, and (ii) influencing the temporal allocation of EV charging load, which indirectly affects power losses and carbon emissions, though these remain smaller in magnitude. Overall, selecting $\rho \approx 0.6$–0.9 provides a balance between foresight and robustness and, when combined with the Adaptive H policy, delivers the lowest cumulative costs observed in the case study.

4. Conclusions

This study proposes an adaptive rolling-horizon optimization framework for coupled transportation and power systems, in which the prediction window is dynamically adjusted in response to the volatility of OD demand and renewable generation. By explicitly balancing the benefits of foresight against the risks of forecast errors, the framework provides a more resilient and adaptive mechanism for multi-period cross-domain operation. In contrast to conventional fixed-horizon or no-forecast approaches, the proposed method is able to regulate the prediction window online, thereby enhancing the system's capability to cope with uncertainty. The numerical experiments further demonstrate that forecasting-based strategies can substantially alleviate downstream congestion and reduce the overall system operating cost, considering both travel time on the transportation side and system losses and carbon emissions on the power side. Moreover, the adaptive horizon consistently outperforms the fixed horizon by maintaining robustness under volatile conditions, confirming the effectiveness and resilience of the approach. These findings suggest

 https://doi.org/10.3390/en19010227

that adaptive horizon control provides a practical pathway for managing the intertwined uncertainties of future low-carbon transportation and power systems.

Future research will extend the proposed framework to larger-scale coupled transportation-power systems and investigate advanced computational strategies to further enhance scalability and efficiency. In addition, future work will relax the perfect-rationality assumption by incorporating stochastic user equilibrium or bounded rationality behavior models to better capture realistic EV routing and charging decisions.

Author Contributions: Conceptualization, Methodology, Writing—Original Draft, and Funding Acquisition, Z.Z.; Formal Analysis, Resources, Project Administration, and Writing—Review and Editing, S.L.; Investigation, Data Curation, and Writing—Review and Editing, Y.W.; Funding Acquisition, Conceptualization, and Writing—Review and Editing, F.T.; Visualization, Investigation, and Writing—Original Draft, H.L.; Software, Investigation, and Writing—Original Draft, P.S.; Supervision, Conceptualization, and Writing—Review and Editing, C.Y. All authors have read and agreed to the published version of the manuscript.

Funding: This work was funded by the Science and Technology Project of State Grid Hebei Electric Power Co., Ltd., with the project title "Research and Application of Integrated Transmission and Distribution Low-Carbon Dispatching Technology for Urban Power Grids Considering Vehicle-Grid Integration" (kj2024-061).

Data Availability Statement: No external data were used in this study. All data presented in this manuscript can be fully reproduced based on the models, parameters, and algorithms described in the paper.

Conflicts of Interest: Authors Zhe Zhang, Shiyan Luan, Yingli Wei, Fan Tang, and Haosen Li were employed by the Xiong'an New Area Power Supply Company, State Grid Hebei Electric Power Co., Ltd. The remaining authors declare that the research was conducted in the absence of any commercial or financial relationships that could be construed as a potential conflict of interest. The authors declare that this study received funding from the Science and Technology Project of State Grid Hebei Electric Power Co., Ltd. (kj2024-061). The funder was not involved in the study design, collection, analysis, interpretation of data, the writing of this article, or the decision to submit it for publication.

References

1. Wu, T.; Zhuang, H.; Huang, Q.; Xia, S.; Zhou, Y.; Gan, W.; Terzić, J.S. Routing and scheduling of mobile energy storage systems in active distribution networks based on probabilistic voltage sensitivity analysis and Hall's theorem. *Appl. Energy* **2025**, *386*, 125535. [CrossRef]
2. Wu, T.; Li, Z.Y.; Wang, G.B.; Zhang, X.; Qiu, J. Low-carbon charging facilities planning for electric vehicles based on a novel travel route choice model. *IEEE Trans. Intell. Transp. Syst.* **2023**, *24*, 5908–5922. [CrossRef]
3. Wang, C.; Liu, C.; Chen, J.; Zhang, G. Cooperative planning of renewable energy generation and multi-timescale flexible resources in active distribution networks. *Appl. Energy* **2024**, *356*, 122429. [CrossRef]
4. Pham, V.H.; Ahn, H.-S. Distributed stochastic model predictive control for an urban traffic network. *IEEE Trans. Intell. Transp. Syst.* **2023**, *early access*.
5. Wu, T.; Yu, H.; Bu, S.; Xia, S.; Wang, H. An improved dynamic user equilibrium model-based planning strategy of fast-charging stations for electric vehicles. *IEEE Trans. Transp. Electrif.* **2025**, *11*, 1555–1569. [CrossRef]
6. Cao, Y.; Zhao, H.; Liang, G.; Zhao, J.; Liao, H.; Yang, C. Fast and explainable warm-start point learning for AC Optimal Power Flow using decision tree. *Int. J. Electr. Power Energy Syst.* **2023**, *153*, 109369. [CrossRef]
7. Zhang, H.; Moura, S.J.; Hu, Z.; Song, Y. PEV fast-charging station siting and sizing on coupled transportation and power networks. *IEEE Trans. Smart Grid* **2018**, *9*, 2595–2605. [CrossRef]
8. Wang, G.; Zhang, X.; Wang, H.; Peng, J.-C.; Jiang, H.; Liu, Y.; Wu, C.; Xu, Z.; Liu, W. Robust planning of electric vehicle charging facilities with an advanced evaluation method. *IEEE Trans. Ind. Inform.* **2018**, *14*, 866–876. [CrossRef]
9. Crozier, C.; Deakin, M.; Morstyn, T.; McCulloch, M. Coordinated electric vehicle charging to reduce losses without network impedances. *IET Smart Grid* **2020**, *3*, 677–685. [CrossRef]

 https://doi.org/10.3390/en19010227

10. Sortomme, E.; Hindi, M.M.; MacPherson, S.D.J.; Venkata, S.S. Coordinated charging of plug-in hybrid electric vehicles to minimize distribution system losses. *IEEE Trans. Smart Grid* **2011**, *2*, 198–205. [CrossRef]
11. Yang, C.; Liang, G.; Liu, J.; Liu, G.; Yang, H.; Zhao, J.; Dong, Z. A non-intrusive carbon emission accounting method for industrial corporations from the perspective of modern power systems. *Appl. Energy* **2023**, *350*, 121712. [CrossRef]
12. Wei, W.; Wu, L.; Wang, J.; Mei, S. Expansion planning of urban electrified transportation networks: A mixed-integer convex programming approach. *IEEE Trans. Transp. Electrif.* **2017**, *3*, 210–224. [CrossRef]
13. Gan, W.; Shahidehpour, M.; Yan, M.; Guo, J.; Yao, W.; Paaso, A.; Zhang, L.; Wen, J. Coordinated planning of transportation and electric power networks with the proliferation of electric vehicles. *IEEE Trans. Smart Grid* **2020**, *11*, 4005–4016. [CrossRef]
14. Duan, X.; Hu, Z.; Song, Y.; Strunz, K.; Cui, Y.; Liu, L. Planning strategy for an electric vehicle fast charging service provider in a competitive environment. *IEEE Trans. Transp. Electrif.* **2022**, *8*, 3056–3067. [CrossRef]
15. Fan, V.H.; Dong, Z.; Meng, K. Integrated distribution expansion planning considering stochastic renewable energy resources and electric vehicles. *Appl. Energy* **2020**, *278*, 115720. [CrossRef]
16. Xi, W.; Chen, Q.; Xu, H.; Xu, Q. A Bi-Level Demand Response Framework Based on Customer Directrix Load for Power Systems with High Renewable Integration. *Energies* **2025**, *18*, 3652. [CrossRef]
17. Alaraj, M.; Radi, M.; Alsisi, E.; Majdalawieh, M.; Darwish, M. Machine Learning-Based Electric Vehicle Charging Demand Forecasting: A Systematized Literature Review. *Energies* **2025**, *18*, 4779. [CrossRef]
18. Yang, C.; Liu, J.; Liao, H.; Liang, G.; Zhao, J. An improved carbon emission flow method for the power grid with prosumers. *Energy Rep.* **2023**, *9*, 114–121. [CrossRef]
19. Wang, H.; Ye, Y.; Wang, Q.; Tang, Y.; Strbac, G. An efficient LP-based approach for spatial–temporal coordination of electric vehicles in electricity–transportation nexus. *IEEE Trans. Power Syst.* **2023**, *38*, 2914–2925. [CrossRef]
20. Xie, S.; Xu, Y.; Zheng, X. On dynamic network equilibrium of a coupled power and transportation network. *IEEE Trans. Smart Grid* **2022**, *13*, 1398–1411. [CrossRef]
21. Wang, X.; Shahidehpour, M.; Jiang, C.; Li, Z. Resilience enhancement strategies for power distribution network coupled with urban transportation system. *IEEE Trans. Smart Grid* **2019**, *10*, 4068–4079. [CrossRef]
22. Zielińska, A.; Jankowski, R. Forecasting Installation Demand Using Machine Learning: Evidence from a Large PV Installer in Poland. *Energies* **2025**, *18*, 4998. [CrossRef]
23. Dörfler, F.; Simpson-Porco, J.W.; Bullo, F. Breaking the hierarchy: Distributed control and economic MPC of AC power systems. *IEEE Trans. Control Netw. Syst.* **2016**, *3*, 241–253. [CrossRef]
24. Stanojev, O.; Markovic, U.; Aristidou, P.; Hug, G.; Callaway, D.; Vrettos, E. MPC-based fast frequency control of voltage source converters in low-inertia power systems. *IEEE Trans. Power Syst.* **2022**, *37*, 3209–3220. [CrossRef]
25. An, Z.; Liu, X.; Xiao, G.; Zhang, M.; Pan, Z.; Kang, Y.; Jenkins, N. Learning-based tube MPC for multi-area interconnected power systems with wind power and HESS: A set identification strategy. *IEEE Trans. Autom. Sci. Eng.* **2025**, *22*, 20458–20468. [CrossRef]
26. Huang, W.; Gao, C.; Li, R.; Bhakar, R.; Tai, N.; Yu, M. A model predictive control-based voltage optimization method for highway transportation power supply networks with soft open points. *IEEE Trans. Ind. Appl.* **2024**, *60*, 1141–1150. [CrossRef]
27. Naseri, F.; Farjah, E.; Kazemi, Z.; Schaltz, E.; Ghanbari, T.; Schanen, J.-L. Dynamic stabilization of DC traction systems using a supercapacitor-based active stabilizer with model predictive control. *IEEE Trans. Transp. Electrif.* **2020**, *6*, 228–240. [CrossRef]
28. United States Bureau of Public Roads. *Traffic Assignment Manual for Application with a Large, High Speed Computer*; U.S. Department of Commerce, Bureau of Public Roads, Office of Planning, Urban Planning Division: Washington, DC, USA, 1964.

 https://doi.org/10.3390/en19010227

MDPI AG
Grosspeteranlage 5
4052 Basel
Switzerland
Tel.: +41 61 683 77 34

Energies Editorial Office
E-mail: energies@mdpi.com
www.mdpi.com/journal/energies

www.ingramcontent.com/pod-product-compliance
Lightning Source LLC
Chambersburg PA
CBHW061603210926
51850CB00009B/2197
9783725870905